Dilemmas of hydropower development in Vietnam: between dam-induced displacement and sustainable development

Những nghịch lý trong phát triển thủy điện ở Việt Nam: di dân do xây dựng đập thủy điện và phát triển bền vững

(với một phần tóm tắt bằng tiếng Việt)

Ty Pham Huu

Eburon

Delft 2015

ISBN 978-90-5972-959-9

Uitgeverij Eburon
Postbus 2867
2601 CW Delft
tel.: 015-2131484 / fax: 015-2146888
info@eburon.nl / www.eburon.nl

Cover design and pictures: Ty Pham Huu
Cartography and design figures: Ty Pham Huu

TABLE OF CONTENTS

LIST OF FIGURES

LIST OF TABLES

LIST OF BOX

ABBREVIATIONS

ADB	Asian Development Bank
ARBCP	Asia Regional Biodiversity Conservation Program
BCAP	Biodiversity Conservation Action Plan
BCSR	Board on Compensation, Support, and Resettlement
CODE	Center of Development and Consultation
COMINGO	Committee for Foreign NGO Affairs
CORENARM	Consultative and Research Centre on Natural Resource Management
CPC	Commune People's Committee
CPI	Consumer Price Index
CRD	Centre for Rural Development in Central Vietnam
CSOs	Civil Society Organizations
CSRD	Centre for social research and development
CT-TW	Circular of Central Communist Party
DARD	Provincial Department of Agriculture and Rural Development
DFID	Department for International Development
DIDR	Development-Induced Displacement and Resettlement
DNWACO	Dong Nai Water Company
DoNRE	Department of Natural Resources and Environment
DPC	District People's Committee
EREN	Energy Efficiency and Renewable Network
EVN	Vietnam Electricity Group
FDI	Foreign Direct Investment
FPD	Forest Protection Department
FPES	Payment for forest environmental services
FPIC	Free, Prior, Informed Consent
FPMU	Forest Protection Management Unit
GDP	Gross Domestic Product
GONGOs	Government-Organized Non-Governmental Organizations
GreenID	Centre for Innovation and Development

GSO	Official Development Assistance
GTZ	German Development Agency
Ha	Hectare
HRPFMB	Huong River Protection Forest Management Board
HUAF	Hue University of Agriculture and Forestry
ICOLD	International Commission on Large Dams
IEA	International Energy Association
IFAD	International Fund for Agricultural Development
IFC	International Finance Corporation
IHA	International Hydropower Association
INGO	International Non-Government Organization
IPCC	Intergovernmental Panel on Climate Change
IRENA	International Renewable Energy Agency
IRN	International River Network
IRR	Impoverishment Risks and Reconstruction
JBIC	Japan Bank for International Cooperation
MARD	Ministry of Agriculture and Rural Development
MDG	Millennium Development Goals
MOF	Ministry of Finance
MOIT	Ministry of Industry and Trade
MONRE	Ministry of Natural Resources and Environment
MPI	Ministry of Planning and Investment
MW	Megawatt
NCEIAA	National Committee on Environmental Impact Assessment and Approval
ND-CP	Decree of Vietnam Government
NGO	Non-Government Organization
NIAAP	Institute of Agriculture Planning and Design
OECD	Organisation for Economic Co-operation
OP	Operational Policy
PACCOM	People's Aid Co-coordinating Committee
PECC1	Power Engineering Consulting Joint Stock Company

PPC	Provincial People's Committee
QD-TTg	Decision of the Prime Minister of Vietnam
QD-UBND	Decision of Provincial People's Committee
RDMA	Regional Development Mission for Asia
REN21	Renewable Energy Policy Network for the 21st Century
SAWACO	Sai Gon water company
SNV	Netherlands Development Organisation
TT-BNNPTNT	Circular of MARD
TWh	Terawatt-hours
UNDESA	Department of Economic and Social Affairs of the United Nations
UNDP	United Nations Development Programme
USEIA	Energy Information Administration of United States
VND	Vietnamese Dong
VNGO	Vietnamese Non-Government Organization
VRN	Vietnam River Network
VUFO	Vietnam Union of Friendship Organizations
VUSTA	Vietnam Union of Science and Technology
WB	World Bank
WCD	World Commission on Dams
WCED	World Commission on Environment and Development
WTO	World Trade Organization

ACKNOWLEDGEMENTS

Looking back the routine of my PhD program, I feel very fortunate to have received a lot of valuable guidance and supports from many people and institutions. Without these supports, I would never have finished my dissertation.

First of all, I would like to thank my supervisors deeply, Annelies Zoomers, Guus van Westen, and Paul Burgers, for your great support. You taught me how to select appropriate ideas and turn these into worthy research. I am always impressed by discussions with Annelies and Guus because both of you inspired me very much to brainstorm and restructure my writings and helped me enhance my papers to a higher level. You also guided me in how to write and publish valuable journal articles. Also, you have given me many opportunities to participate in IDS research training and activities. These activities have helped me gain more practical knowledge and skills while in the Netherlands. I will never forget what you taught me in doing research. I will bring these good experiences as a means to contribute to the research and scholarship of my university in Vietnam. Guus, I have learnt a lot from your specific editing for my paper articles; it was really helpful for me to write better. I am also very happy that you always took care of me and my Vietnamese PhD fellows in IDS. Thank you so much for your kindness and enthusiasm in helping us. Paul, I am very grateful for your valuable support for my PhD program in IDS. You always cheered me up and gave me valuable comments for my research proposal. It was a great pleasure and honor to work with all of you.

I also want to express my thanks to the University of Utrecht, the Faculty of Geosciences, and the IDS group for your supports. You provided me with favorable conditions for my learning and research in the Netherlands. I accessed a very good education system with full equipment support and effective management system. I am really grateful for the dedicated care and efficiency of the staffs of the Faculty of Geosciences. I felt very comfortable and happy to work with professors and colleagues in IDS—Paul, Maggi, Annelet, Henk, Gery, Caroline, and Gemma. Especially, I have unforgettable memories with PhD fellows in IDS—Femke, Ari, Dinu, Michelle, Suseno, Joseph, Antony, Lucia, Claver, Ignace, Alda, George, Tu, Phuc, and Rizki. Your contribution during discussions of Writers Group for my article papers was also very helpful for me to improve the papers better and faster. I will never forget the time I spent with all of you. My life in the Netherlands was really warm and happy when I had great opportunities to integrate into the family atmosphere of the PhD fellows in the IDS group.

I would like to thank Hue University, Hue College of Agriculture and Forestry, Faculty of Land Resources and Agricultural Environment, which provided me with favorable conditions throughout the PhD program in the Netherlands. I especially thank very much my young colleagues in the Faculty of Land Resources and Agricultural Environment for supporting and effectively cooperating with me in collecting data for my research projects. You were always enthusiastic and ready to arrange your valuable time for long trips to the resettlement areas where often remotely mountainous regions are. I am very lucky to have your supports and collaboration during my field work in Vietnam.

I am really grateful to the people of resettlement villages, Bo Hon and Kan Tom in Thua Thien Hue province who enthusiastically shared their precious time for interviews and provided valuable information for my research project. Also, I would like to thank the leaders and staffs of the local authorities in Huong Tra, A luoi, Dong Giang district, staff of Binh Dien, A luoi, and A Vuong hydropower plants, staff of the A vuong protection forest management board for sharing time and efforts to provide me valuable information and suggestions. Furthermore, I am very grateful to the non-governmental organizations (NGOs) in Vietnam for sharing information on their activities related to the theme of my research. I am really impressed by the enthusiastic dedication of Vietnamese NGOs: Vietnam River Network, Centre for Social Research and Development (CSRD), Consultative and Research Centre on Natural Resources Management (CORENARM), Green Innovation and Development Centre (GreenID), and the contribution of international NGOs when it came to their participation in improving the quality of dam-induced displacement and resettlement policies, advocating for displaced communities, and formulating benefit sharing policies from hydropower development. In particular, I am very impressed with the great dedication of Ms. Lam Thi Thu Suu, the Director of CRSD, who has coordinated VNR and other Vietnamese NGOs to advocate to displaced communities. She received the well-deserved Rockefeller Foundation Bellagio Centre award in 2013 for her NGO leadership in Vietnam. It was a great honor to have your information sharing related to your activities and achievements. I am also very grateful and honor to have discussions with Prof. Dang Hung Vo, a former Vice Minister of Ministry of Natural Resources and Environment (MONRE) of Vietnam. Your sharing helped me find out the most important reason why the problem of land acquisition, compensation, displacement and resettlement exists in Vietnam.

I am also very thankful to the Netherlands Organization for International Cooperation in Higher Education (NUFFIC), the Vietnam International Education Development (VIED), and the IS Academy on Land Governance for Equitable and Sustainable Development (LandAC) for securing funding for my PhD program in the Netherlands. I would like to thank Associate Prof. Jane Singer of Kyoto University for teaming up with me to organize field trips and workshops and for the articles we wrote together related to the topic of dam-induced displacement and resettlement. I would like to thank and am highly appreciative of Dr. Bill Hart, the former supervisor of my Master's program at Dalhousie University, Canada, who had an unconditional enthusiasm in reading the chapters and editing the manuscript. I also appreciate Brian Hotson, MTS, director of Writing Centre at Saint Mary's University, Canada, who spent his time and effort to complete English editing of the manuscript. I am also very appreciative for the support of many people in the Van Unnik building and Utrecht University, the secretariat, security staff, and officials of human resource, finance, ICT, Pedel, URU, and many others.

My dear friends, I want to sincerely thank you for your follow-up and encouragement throughout my PhD journey. You always gave me your enthusiastic and forward-looking support for good things for me, and I am really happy to have you around. Tú, I always owe you for what you did for me when I applied for the PhD program in IDS. Phục, I thank you for always encouraging and sharing your ideas and social life as we worked together in IDS. I wish both of you luck and success always. Also, I would like to thank all the Vietnamese friends in the Netherlands, Huy, Nam, Nam, Hien, Hang, Thuy, Ha Anh, Trang, Trang, Nhung, Thanh, Kim Anh, Ha Anh, Giang, Tuan, An, and many others. You always received me with warm hearts as fellow countrymen living in the beautiful country, the Netherlands. Having all of you around made me feel at home and stronger on the journey of my PhD program. I also thank Leo Horowitz, who is always ready to give me a chance to stay at his house in the Netherlands. I wish you all healthy and successful.

I want to express my deepest thanks to my family, my parents, brothers and sisters, and everyone in the family, who always said "yes" enthusiastically for the learning path of mine and the care for my wife and my daughters during my time in the Netherlands. Last but not least, I would like to tell my wife, Lý, and my daughters, Ngọc Khánh and Minh Ngọc, that I love all of you very much because you always comfort me, inspire me, and cheer me up. Thank you for your love and patience.

Utrecht, October 2014

1   INTRODUCTION

1.1   Rationale

In recent years, hydropower has become an issue in Vietnamese society, especially in political forums, mass media, academia, social and environmental movements, and daily conversation. Many people from diverse sectors of Vietnamese society have raised questions of whether hydropower is a clean and sustainable energy because of its adverse consequences on society and environment. These questions focus on suspicions of aspects of hydropower's sustainability and equity. Many perceive the benefits of hydropower dam construction and support further construction; whereas, others assert that Vietnam should stop hydropower dam construction because these are risky for society and environment. I witnessed this contrasting opinion in an annual conference of the Vietnam River Network (VRN) in 2012, where many environmental and social activists and researchers were critical of dam construction.

> Many people believe benefits of hydropower are large and we must accept the loss of environmental and ecological value for economic development, by contrast we believe that hydropower dams have caused tremendously adverse consequences to environment and society and therefore we should stop building more hydropower dams in Vietnam. (VRN, 2012)

However, representatives of Ministry of Natural Resources and Environment (MONRE) and Ministry of Industry and Trade (MOIT) proposed that hydropower enabled national economic growth, and that the contribution of the private sector in dam construction was important for national development. Therefore, MONRE and MOIT asked environmental and social activists to support hydropower companies to advance hydropower dam construction for the sake of national socio-economic development. However, there was no agreement. MONRE and MOIT representatives left the conference without notice to conference organizers and discontinued the debates of the conference (Author's notes, 2012). In addition, I have participated in similar conferences organized from national to local levels where divergent opinions were expressed between social and environmental activists and government officials regarding hydropower development in Vietnam. Debate is on-going without a foreseeable end. The debate often revolves around issues of accountability and state agency and investor responsiveness in addressing environmental and social impacts of dam construction, particularly in dealing with problems of displaced people. Policies of compensation, displacement and resettlement, livelihood supports, and benefit-sharing for affected people are always central attentions in these debates.

The aim of this dissertation is to explain why hydropower development has emerged as a developmental issue of contemporary Vietnam and its implications for equitable and sustainable development. In particularly, we will take a closer look at hydropower dam construction costs and benefits and examine how hydropower has been initiated and implemented to seek solutions for better governance, where the roles and influences of different actors in hydropower development, including government, local authorities, private sector, civil society organizations, international development agencies, affected communities, are clarified from national to local levels to evaluate the governance of hydropower development in Vietnam.

As hydropower development is a globally controversial issue, I will provide background to this research by characterizing discourses and trends of different global stakeholders regarding the pros, cons, and governance matters of hydropower dam development.

1.2   Global debates on hydropower

Before 1975, hydropower dam construction was considered a symbol of nation-building and national pride for the growth of many developed countries, but this perception changed significantly as a result of environmental and social movements. Advocacy of social and environmental activists have changed mind-sets regarding hydropower development where now a majority of people pay more attention to environmental and social values rather than economic values regarding hydropower dam construction decision-making, especially those in developed countries (Biswas, 2012). Hydropower renewability and sustainability has been questioned worldwide (Gagnon et al., 2002; Wang et al., 2014). The controversy has polarized into two camps: opponents and proponents to hydropower dam construction.

Opponents strongly be in opposition to hydropower dam construction (Biswas, 2012; Nursue, 2003) and criticize the poor performance of hydropower dams (WCD, 2000; McCully, 2001; Scudder, 2005; Sigh, 2002; Savacool & Bulan, 2011; Biswas, 2012; Wang et al., 2014; IRN, 2014). They condemn hydropower dams because they cause huge environmental and social costs that far outweigh any benefit dams provide to society (McCully, 2001; Stone, 2011; WCD, 2000). For example, the cost for mitigating environmental degradation in the case of Three Gorges hydroelectric is estimated at $26.45 billion in the next ten years. Also, cost overrun is high, about 45%, and, in some cases, more than 200%, leading to delays in commissioning projects that affect the delivery of services, increasing interest payments, and delaying revenue generation (WCD, 2000, p. 41). In addition, the intended benefits of hydropower dams are uncertain (Stone, 2011, p. 817; Ansar et al.,

2

2014). Also, opponents deny that hydropower is non-greenhouse gas emission energy because hydropower dams also release a certain amount of greenhouse gas (GHG). The gross emission of GHG may account for between 1% and 28% of the global warming potential (WCD, 2000, p. 75). Another important criticism is that hydropower dams cause a tremendous loss of land and displace large populations, often becoming worse off than before displacement. According to the World Commission on Dams (2000) and ICOLD (2010), a total of 9,423 hydropower dams cover 500,000 km², displacing approximately10 million people globally (Cernea, 2000). Opponents also argue that hydroelectric projects only benefit the rich, while displaced peoples' living conditions become worse after resettlement. Benefits from producing hydroelectricity often go to the pockets of construction companies, engineering consultancy companies, corruption politicians, and government officials; whereas, displaced people bear most of the negative impacts from dam construction (Biswas, 2012; Wang et al., 2014).

In reaction to this criticism, proponents of hydropower development often argue that hydropower is the most mature, reliable and flexible, available, renewable, and cost-effective electricity generation technology. Hydropower plants typically operate at an electrical efficiency[1] of 85% to 95% (EREN, 2014). This compares to about 55% for combined-cycle gas turbines and 30% for wind power (Japan Society of Energy and Resources, 1996), 30 % to 40% for coal or oil fired plants (IEA, 1990), and 7% to 17% for solar photovoltaic panels (EREN, 2014). Hydropower plants usually have very long lifespan, with a range of 30 to 80 years (Brown, 2011; Mitigation, 2011). Hydropower is the second load following power capability[2] after gas turbine power plants (IRENA, 2012). Hydroelectric generating units are able to start up quickly and operate efficiently almost instantly. This is opposite to thermal plant where start-up can take several hours or more (Brown, 2011). Furthermore, hydropower is the only large-scale and cost-efficient storage technology available today (IEA, 2010). It can serve as a power source for large and small as well as centralized and isolated grids. Small hydropower dams can be a cost-competitive option for rural electrification for remote communities in developed and developing countries.

Another advantage of hydropower technology in developing countries is that it can have an important multiplier effects by providing both energy and water supply services (e.g., flood control and irrigation), thus bringing social and economic benefits (IRENA, 2012). Among energy sources, hydropower has the highest energy payback

---

[1] Electrical efficiency is defined as useful power output divided by the total electrical power consumed.

[2] Load following power capacity is the ability of a power plant to be able to adjust power production as demand for daily electricity fluctuation.

(IEA, 2000; Jia et al., 2012). More importantly, hydropower has been considered an attractive renewable energy source with the advantage of being less harmful in terms of GHG emissions compared to fossil fuels (IEA, 2008). It plays an important role in reducing global GHG emissions by 10% per annum (WCD, 2000). In its current role, according 2012 statistics, hydropower offsets 6.4 million barrels of oil-equivalent (thermal electric generation) or 1848 Mt of coal or 27,247 million Mcf^3 of natural gas to generate the same amount of electricity (EIA, 2012, p. 182). Therefore, hydropower represents an alternative source of electricity over fossil fuels, the effects of which are the main sources of global warming (Stern, 2008; Bakis, 2007; Jia et al., 2012; Vandal, 2012) because CO_2 concentrations are due primarily to fossil fuel use, making up 56% of CO_2 emissions created by human activities (IPCC, 2007). Out of the 2,062 projects registered with the Clean Development Mechanism (CDM) Executive Board by 1 March 2010, 562 projects were hydropower. When considering the predicted volumes of Certified Emission Reductions to be delivered, registered hydropower projects were expected to avoid more than 50 Mt of carbon dioxide (CO_2) emissions per year by 2012 (Mitigation, 2011).

Beyond its role in contributing to energy supply security and reducing a country's dependence on fossil fuels, hydropower offers opportunities for poverty alleviation. It can contribute to regional cooperation, as good practices in managing water resources require a river basin approach for trans-boundary watersheds. In addition, multipurpose hydropower dams can strengthen a country's ability to adapt to climate change-induced hydrological variability (World Bank, 2009). It also creates job opportunities for workers during construction phase. For example, construction of Kariba and Grand Coulee employed between 10,000 and 15,000 workers each. During the peak construction period, a labour force of about 15,000 people was employed at Tarbela (WCD, 2000).

Although, there are many controversies and anti-dam movements, most international organizations still recognise hydropower as an important energy for eradicating poverty, changing unsustainable patterns of consumption and production, protecting and managing the natural resource base, and sustainably developing the world's least developed countries as stated in the Millennium Development Goals (MDG) since 2000 (The Assembly of United Nations, 2000). The World Summit on Sustainable Development in 2002 concluded that hydropower should be incorporated in the course of development as a renewable energy globally because it does not block the capability of future generations to meet their needs. Hydropower has been developed in its modern form in the last hundred years mainly to provide cheap

---

^{3} Mcf unit equals the volume of 1,000 cubic feet (cf) of natural gas. See
http://www.eia.gov/tools/faqs/faq.cfm?id=667&t=6 for more related units with Mcf

electricity for economic development and energy security. In recent years, hydropower also has been complemented by its potential to the climate change mitigation and adaptation. The Third World Water Forum in 2003 concluded that hydropower is one of the renewable and clean energy sources and its potential should be realized in an environmentally sustainable and socially equitable manner, and a necessity for growth and poverty reduction (World Water Council, 2003, p. 111-29; Biswas, 2012). The main concern is how hydropower projects should be best selected, built, and operated for sustainability, and how past lessons learnt could be used to deal with negative environmental and social impacts towards an environmentally sustainable and socially equitable development (Gagnon, 2002; Korch, 2002; Biswas, 2012, Wang et al., 2013).

For achieving this goal, several important organizations representing both opponents and proponents of hydropower dam construction initiated multi-stakeholder forums in early 2000s on a global scale to formulate standards and guidelines for a global decision-making framework on hydropower dam construction. Two of the most influential associations, including International Environmental Association (EIA) and World Commission on Dams (WCD), proposed two mechanisms. In 1999, IEA proposed a sector-oriented hydropower development standards and guidelines: "*Hydropower and the environment: present context and guidelines for future action*" with five recommendations[4]. In early 2000s, WCD recommended the risks and rights-based approach: "*Dams and Development: a new framework for decision-making*" with seven priorities and 26 principles[5]. There are several similarities between the two as they point out the necessity of evaluating the best options of hydropower dams that must be accepted by the public. Benefits from hydropower development must be shared to ensure that affected people are fairly treated and adequately compensated. However, they diverge because of different conceptions about the development of hydropower. The guidelines of WCD require governments and developers to carry out a lengthy negotiation with all relevant stakeholders to get a complete consensus that could delay projects indefinitely. By contrast, the guidelines of IEA emphasize the enhancement of government responsibilities in setting up an effective and efficient decision-making processes, which could avoid endless negotiations. The guidelines of WCD tend to give a stronger role to international

---

[4] The guidelines of IEA (2000) focuses on supporting hydropower professional development companies, associations, and governments to promote hydropower development in ways that should enhance five ethnical principles, including stewardship (responsible management); participatory decision making; farsightedness and control; fairness and justice; optimality.

[5] The new decision making framework of WCD (2000),which takes the rights and risks-based approach setting the decision-making in the form of the seven strategic priorities, including: gaining public acceptance; comprehensive options assessment; addressing existing dams; sustaining rivers and livelihoods; recognizing entitlements and sharing, benefits; ensuring compliance; sharing rivers for peace, development and security.

organizations, especially anti-dam NGOs, whereas those of IEA aim to support hydropower developers to continue promoting hydropower in ways of improving the decision-making process. WCD assumes that other renewable energy options, such as wind power, could realistically replace, energy services provided by hydropower dams, whereas IEA considers coal-based electricity generation as the main competitor of hydropower, and, therefore, any limits placed on hydropower development would result in severe, negative environmental impacts due to an increase in GHG emission from burning coal (Gagnon, 2002).

Since the proposal of the two decision-making models for dam building, many scholars have given their viewpoints that are either supportive or critical. Continuing debates have become emotionally and dogmatic, but it is crucial in the search for best solutions to solve problems of hydropower development. However, Biswas (2012) noted that:

> No single solution is the most appropriate for all countries at any specific time, or even in the same country because of differing climatic, technical, economic, social or institutional conditions. Water development patterns will vary with time, knowledge and experience, as will development paradigms. Nothing is permanent. Countries are often at different stages of development. No two countries are identical. Their economic and management capacities are not identical; climatic, physical and environmental conditions are often dissimilar; institutional and legal frameworks for water management differ; and, social and cultural conditions vary significantly. (p. 17)

The sustainability of hydropower also largely depends on the specific governance mechanism of each country. If the trade-offs amongst poverty reduction, economic growth, income redistribution and environmental protection are governed objectively, accurately, honestly, sensitively, and in a socially acceptable manner at all levels and scales, best solutions could be sought for each specific hydropower development case in sustainable manners (Biwas, 2012; Gagnon, 2002).

Thus, it can be seen that there is considerable attention paid for formulating global principles and guidelines for proper governance, but there is less of an eye for differences in specific local and national settings that affect the way in which such global guidelines can work successfully. Therefore, it is essential to understand how hydropower is initiated, decided, implemented, and monitored in different countries where various actors with different powers play diverse roles and have different influences on the hydropower development in a specific country.

1.3 Research design

1.3.1 Research questions

Provided that the main goal of this study was to strengthen our understanding of hydropower development and to seek implications for equitable and sustainable development in Vietnam, and, therefore, this research seeks to answer the following, central question:

Under what conditions can hydropower dam development contribute to equitable and sustainable development?

This question was posed as a means to search for better governance to address the sustainability problems of hydropower development. The literature review indicates that many international standards, guidelines, and governance norms have been introduced to less developed countries, but they were not strongly accepted by countries in which these were introduced. This proves that the principle, *"think globally, act locally,"* is not really effective in the context of the proposed model of global hydropower governance, as the nature of specific local and national contexts is not known and fixed. This "reception fault" does matter for the successful implementation of global decision-making framework as argued on pages from 11 to 14. Therefore, lessons learnt from the case of Vietnam will be useful as a means to draw implications for addressing the current problems on hydropower development. In other words, this study will supplement another principle, *"think locally, and act globally,"* to fill the gaps of the global debate about the influences of dissimilar factors of each country, region, locality, affected community, and types of hydropower dam. The answer to the question could be used for the government of Vietnam and local authorities, developing countries, development banks, and international associations to adjust their strategies and to make their actions more feasible and effective to contribute to the sustainability of hydropower development on global, national, and local scales. To answer the central question, five sub-questions are formulated as follows.

1. What are the drivers and characteristics of hydropower development in Vietnam?

This first sub-question intends to identify the drivers of promoting hydropower and to review the progress of hydropower development, benefits, and costs induced by hydropower dam construction in Vietnam.

2. How have hydropower dams been initiated, decided, implemented, operated, and monitored in Vietnam?

This second sub-question aims to identify gaps between the policies and practical implementation regarding to compulsory land acquisition, compensation, displacement and resettlement works, and benefit-sharing mechanisms in Vietnam. The evolution of these policies is reviewed to evaluate whether there are any improvement in terms of time and scale. These policies are investigated from national to local levels to identify problems of implementation and to explain why these problems have occurred.

3. To what extent have people been displaced and influenced because of hydropower dam construction in Vietnam and how did they respond to displacement and resettlement?

The goal of this third sub-question is to identify the impacts of hydropower dam construction on people displaced by hydropower dam construction and to examine how they respond to reconstructing their livelihoods after resettlement. Furthermore, it aims to explain why their livelihoods have been changed as a result of resettlement. The inequity of development within displaced communities is also investigated. Moreover, this question is intended to investigate whether alternatives exist to support displaced people to have better livelihood restoration and sustainable development.

4. What are the roles and influences of NGOs in the hydropower development in Vietnam?

This fourth sub-question is formulated to investigate how NGOs involved in and influenced the decision-making process of hydropower development in Vietnam, and what we could learn from the interaction between NGOs and the state when it comes to addressing the problems of hydropower dam construction in Vietnam.

To answer these questions, first of all, we need an understanding of the existing knowledge and methods to accomplish surveys and analyses. The following conceptual framework, study approaches, and research activities will anchor platforms for this study.

1.3.2 Conceptual framework

In this study, I focus on analyzing the roles of governance in promoting the equitable and sustainable hydropower development, and, therefore, I explain the main concepts related to sustainable development, equity, and good governance. Firstly, sustainable development is defined as "development that meets the needs of the present without compromising the ability of future generations to meet their own needs" (WCED, 1987, p. 8). Here, sustainability is operationalized into three elements and often depicted as diagrams: three concentric circles, three pillars supporting a roof, and three integrated circles (see Figure 1-1). The notion of the three concentric circles is that both social and economic developments are seen to be constrained by the environment. The three pillars of sustainability means that economic, environmental, and social dimensions should be treated equally. The new concept is an interaction between the social, economic and environmental aspects of human activity needs to maintain a healthy environment, promote investment in sustainable livelihoods and support economic growth. This balance is best shown by integrating three circles, which means that any single sustainable dimension must compromise with two other sustainable aspects (Schumann et al., 2010).

Furthermore, sustainability is divided into two types in terms of priority put on each dimension, including strong and weak sustainability. The strong sustainability approach is radical, since it aims to ensure that the pursuit of development must not cause negative impacts on the environment and society, ecosystems should not be changed, and the resource should be not depleted. On the other hand, the weak sustainability approach admits that the pursuit of socio-economic development, especially development programs in poor and underdeveloped countries, cannot be realized without modification of ecosystems and natural resources. And, therefore, the collaboration and sensible compromise is crucial to consider the trade-offs among three dimensions of sustainable development (Beckerman, 1995; Neumayer, 2003).

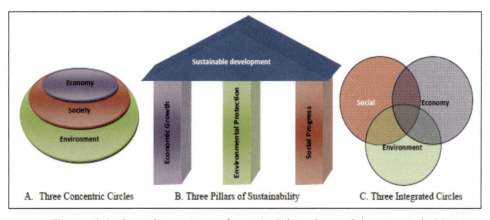

Figure 1-1 Three dimensions of sustainability. *Source: Schumann et al., 2010*

However, the trade-offs among three dimensions of sustainability rely much on the power relation among decision makers. Therefore, it is crucial to raise five questions on sustainable development about: Who initiates?; Who decides?; Who participates?; Who benefits?; and Who controls? If the focus of these questions is the people who are directly influenced by the development, then it can be assured of sustainability (Ecoforum, 1991). However, the power imbalance between parties involved in the decision making requires each to respect the two principles of equity: justice to current and future generations in relation to resource and intergenerational equity; and fairness in sharing of resources to the competing interests at the time of resource development

9

(Jabareen, 2008). This study focuses the second principle of equity, the benefit sharing which requires a balance between economic development of hydropower and its benefit redistribution of development outcomes to affected people.

Equitable development is also consistent with the people-centred development approach proposed by Korten (1984) and later declared by the Manila Declaration on People's Participation and Sustainable Development in 1989. The people-centred approach requires incorporating the values of justice, sustainability, and inclusiveness into decision-making processes. The Manila Declaration (1989) stated that the people-centred development is the only way to achieve sustainable communities and called for the establishment of self-supporting social and economic systems as key elements of a sustainable society. In the people-centred development approach, central elements of participation include democratic processes, government accountability, access to relevant information, and gender equality and elements of justice comprise local ownership, sovereignty of the people and government enablement, employment and income generation (Declaration, 1989; OECD, 1996; Korten, 1990). The principle of people-centred development also parallels with the inclusive development approach. The UN stresses that many people are marginalized from growth because of their gender, ethnicity, age, sexual orientation, and disability. Development can be inclusive only if all groups of people contribute to creating opportunities, share the benefits of development, and participate in decision-making. Therefore, UNDP's human development approach recommends inclusive development, which integrates the standards and principles of human rights, including participation, non-discrimination and accountability (UNDP, 2000).

It is clear that every society desires for justice and sustainable development, but it seems to be very difficult to achieve in practice. One reason is the obstruction of interest groups who dominate decision-making processes and development-implementation activities. Therefore, equitable development can only be achieved when governance is well-designed and effectively implemented for each developmental type at different scales to control power relation and distribution among different stakeholders. UNDP (1997) defined that governance is the exercise of economic, political, and administrative authority to manage a country's affairs at all levels. It places emphasis on the mechanism, processes, and institutions through which citizens and groups convey their interests, perform their legitimate rights, satisfy their involvement, rule of law, responsiveness, consensus orientation, equity, effectiveness and efficiency, accountability and strategic vision. If these aspects are well-organized, good governance is achieved and can contribute significantly to long-term, sustainable development strategies; to policy consistency through vertical and horizontal organization; to an open, transparent procedure, involving and consulting stakeholders; and to bringing sustainable development strategies closer to local communities and to people. Good governance stimulates accountability, transparency, efficiency, and rule of law at all levels and allows well-organized management of human, natural, economic, and financial capital for equitable and sustainable growth, as well as allows for assurances of the participation of civil society organizations (CSOs) in decision-making processes. Good governance and sustainable development are two separate concepts that are definitely tied together. Good governance does not guarantee sustainable development, but its absence severely limits sustainable development and can, at worst, impede sustainability of development (Kardos, 2012).

1.3.3 Research approaches: multi-stakeholder, multi-level, multi-dimensional, and longitudinal approach

Past and current debates indicate that hydropower is a complex and dynamic development encompassing a wide spectrum of financial institutions, project owners, operators, developers, governments, regulatory bodies, intergovernmental agencies, civil society, and affected stakeholders (Schumann et al., 2010). Therefore, it is crucial to employ appropriate approaches to investigate the complexity and dynamics of governance for the case of hydropower development at different levels, scales, stakeholders, and dimensions. To this end, the multi-stakeholder approach is used to examine roles, influences, and interactions of different stakeholders in hydropower development in which nine elements of good governance (Table 1-1) are selected to investigate to what extent good governance was put into practice. Additionally, the multi-level approach is used to address the complexities of multi-scale and multi-sector issues related to hydropower development (Termeer et al., 2010). Hydropower is also closely interlinked with social, environmental, economic, and political issues that require employing the multi-dimensional approach to examine the trade-offs of economic, social, and environmental dimensions and conflicts that have occurred in hydropower development of Vietnam.

At community and household levels, the impacts of displacement and resettlement on livelihoods often take place over a long period of time. Any short-term evaluation will only grasp short-term impacts that do not reflect long-term changes. Also, it is necessary to look at the interaction of displaced people in long-term with upward institutions (commune, district, province, and government), outwards to civil society organizations, international NGOs, and other non-state actors, and within community. Therefore, the longitudinal approach should be employed to quantify and analyze the long-term changes of displaced people due to hydropower dam construction. To carry out this long-term assessment and analysis, we used the revised framework of displacement, sustainable livelihoods and impoverishment risks that McDowell (2002) proposed to investigate the impacts of

11

displacement and resettlement due to hydropower dam construction on displaced communities (See Figure 1-2).

Table 1-1 Elements of good governance. *Source: Adapted from good governance policies of UNDP (1997); World Bank (1994); ADB (2004); Nwanze & Kouka, 2010; WCD (2000); and IEA (2000).*

Elements of governance	Definitions and criteria
Participation	Both directly and indirectly affected persons, men and women, should have a voice in decision-making, either directly or through legitimate, intermediate institutions that represent their interests. There should be a consultation process
Rule of law	Legal frameworks should be fair, predictable, stable, and enforced impartially, particularly with regards to laws on human rights
Transparency	Processes, institutions, and information should be directly accessible to those concerned and affected, and enough information should be provided to render these understandable and controllable
Responsiveness	Institutions and processes should serve all stakeholders
Consensus orientation	Good governance should mediate differing interests in order to reach broad consensus on the best interests of the group and, where possible, on policies and procedures.
Equity	All men and women, directly and indirectly affected people, should have equal opportunity to maintain or improve their well-being
Effectiveness and efficiency	Processes and institutions should produce results that meet needs while making best use of resources
Accountability	Decision-makers in government, the private sector, and civil-society organizations should be accountable to the public, as well as to institutional stakeholders. This accountability differs depending on the organization and whether the decision is internal or external to an organization.
Strategic vision	Leaders and the public should have a broad and long-term perspective on good governance and human development, together with a sense of what is needed for such development.

The revised framework was combined between the sustainable livelihood framework (Scoones, 1998; DFID, 2001; Ellis, 2000; Carney, 2003) and the impoverishment risks model (Cernea, 1997; Cernea & McDowell, 2000). The central analysis of the framework pays high attention to understand the institutional process in resettlers' adaptation strategies that forced displacement dismantles patterns of social organizations and displaced people to reform to confront new challenges. McDowell (2002) also emphasises questions of how different people gain access to different resources at different times; what they do with that access; who is barred entry and who sets the rules; and what are the implications of this for the management of the environment and the pursuit of sustainability. He recognizes that institutions can be advantageous, functioning as gates opening onto opportunities for some in the community, but, equally, institutions can be oppressive, denying agency and access for certain members of the community.

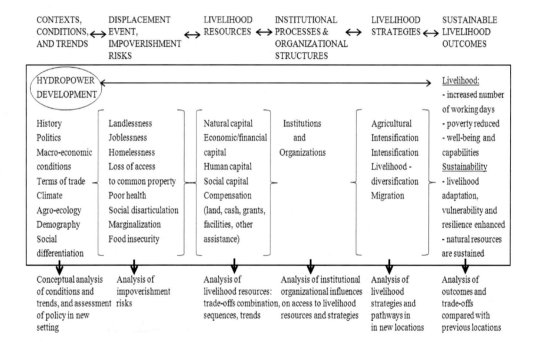

Figure 1-2 Displacement, sustainable livelihoods and impoverishment risks
– the revised model for analysis. *Source: McDowell, 2002*

Cernea (1997) stresses that enabling the rebirth of community institutions is paramount for successful resettlement and livelihood reconstruction. The arrangement of community institutions influences how people access livelihood opportunities (De Haan & Zoomers, 2005). De Haan (2000) considered access and claim as social capital because they embed institutions, relationships, attitudes, and values that govern interaction among people and contribute to economic and social development. Leach at el. (1997) supposed that access and control of natural resources are socially differentiated, in which a set of interacting and overlapping institutions embedded in the political and social life of the area mediate access and control of natural resources. Additionally, social actors are significantly influenced by how institutional arrangements and ecologies vary for various parts of the landscape. Ribot & Peluso (2003) defined that "access is the ability to derive benefit from things"; they emphasize the notion of bundles of powers than that of rights, and include a wider range of social relationships that constraint or enable the benefit from resource use. In the process of displacement and resettlement, for example, hydropower development-induced displacement, institutional arrangements are fundamentally altered and new dynamics influence people's access to and control over resources. The power relation is an important factor that influences displaced people's ability to recover from development-induced shocks.

1.3.4 Research site

Vietnam was selected for this study because the country has been actively promoting hydropower dam construction for a long time after independence in 1945. The country has nearly exploited its hydropower potential and will have accomplished the plan of hydropower dam construction by 2030. In Vietnam, the main goal of promoting hydroelectric development is to ensure electricity security for the industrialization and modernization policies to 2020. The government has put the policy *"electricity must go earlier than other policies"* on the top of development agenda of the country to date (EVN, 2009). Also, the government has a strong commitment to equitable and sustainable development as the following statements in Decision 153/QD-TTg (2004) and Decision 432/QD-TTg (2012) of the Prime Minister of Vietnam on the Sustainable Development Strategy:

> Human beings are the center of sustainable development. Promote the role of people as the key subject, resources and targets of sustainable development; increasingly meet the material and spiritual demand of people of all social classes; build a wealthy and strong country, democratic, equal and civilized society; develop an independent and self-reliant economy with active international integration for sustainable development.(Article 1, Decision 153/QD-TTg)

and

> Sustainable development is the common work of the whole Party, entire people, authorities at all levels, ministries, agencies, localities, enterprises, social organizations, communities and individuals.(Decision 432/QD-TTg, 2012, p. 60)

Moreover, Vietnam has promoted many specific strategies to sustainable development. Since 1986, Vietnam has implemented many economic reforms with notable achievements in economic growth and poverty reduction. The average annual GDP growth rate was 7.85% between 2000 and 2008. There was a three-fold increase in GDP per capita between 2000 and 2011, from US$400 in 2000 to US$1,200 in 2011. This increase has helped Vietnam become a middle-income country. The poverty rates have dropped noticeably, from 28.9% in 2002 to 9.45% in 2010 with 2% reduction per year nationwide. Furthermore, Vietnam accomplished the universalization target of primary education in 2000. Many environmental protection policies have been formulated to promote green growth. The system of state agencies on environmental protection has been established from the central to local levels (Government of Vietnam, 2012). However, Vietnam is still facing many challenges to achieve sustainable development, such as corruption, environmental pollution, inequality, increased poor-rich gap, debt, banking and real estate crisis, rapidly increased prices and high inflation, social problems, traffic accidents, natural calamity and climate change impacts, epidemics, and other problems.

In Vietnam, a single party rules the country as regulated by the new Constitutional 2013. Under this system, the political regime is the Party is the leader, the state is the manager, and the entire people are the owner (National Assembly of Vietnam, 2013). The party organizations are formed from central, provincial, district, commune, and to village level. According to the Constitution 2013 of Vietnam, the state is organized in four systems, including the National Assembly and People's Councils at province, district, and commune level (from the people directly elected); the Government, ministries, ministerial-level agencies, agencies attached to the Government, and People's Committees at province, district, and commune level; the Supreme People's Court and People's courts at province and district level; the Supreme People's Procuracy of Vietnam and the People's Procuracy at province and district level. The ownership of the entire people is institutionalized in the grassroots policy *"people know, people discuss, people do and people check"* (Communist Party of Vietnam, 2006). This policy aims at enhancing the participation of citizens in decision-making on the development of the country and localities, and to promote grassroots democracy. Under this political system, the governance of hydropower development in Vietnam definitely embeds different characteristics compared to other countries. Therefore, the findings of hydropower development in Vietnam will provide useful lessons to global hydropower governance models and for other developing countries accelerating hydropower dam construction.

1.3.5   Methods and research activities

In this study, both qualitative and quantitative methods were used to collect data from different stakeholders and to quantify livelihood impacts of displacement and resettlement on displaced communities, respectively. To increase the credibility and validity of the results, we applied triangulation approaches for collecting data. We focused on data and methodological triangulation presented in Hennink et al., (2011). We collected both secondary and primary data for analysis.

Secondary data was collected from sources such as international organizations, the government of Vietnam, local authorities, hydropower companies, mass organizations, local and international NGOs, research and consultancy institutes and centres, state companies, mass media, and other organizations. Secondary data includes metadata, legal documents, guidelines, scholarly articles, online newspapers, reports, conference proceedings, statistical documents, and other unpublished documents. The secondary data was often employed to contextualize existing policies, guidelines, standards, and implementation of compulsory land acquisition, compensation, displacement and resettlement, livelihood reconstruction, and benefit sharing from global to local levels, from government to non-governmental actors. From that, field work for collecting primary data were designed and implemented.

This primary data collection was carried out for four years between 2010 and 2013. First, **key informant interviews** were carried out with leaders and staff of government, relevant ministries, provinces, districts, communes, villages, hydropower companies, research and consultancy centres, and INGOs and VNGOs to acquire information about their roles, actions, and opinions about issues related to hydropower development that they have been involved in. After that, **focus group discussions** were organized with displaced people to get more information about their communities before and after displacement, to understand how and to what extent they were involved in decision-making on land acquisition, compensation, displacement and resettlement, livelihood reconstruction, and benefit sharing. Each group included 8 to 12 persons, including leaders of villages, patriarch, elders, and selected representatives of other households. Focus group discussions initially also helped identify emerging issues that displaced communities faced with during and after displacement in order to cross-check information gathered from key informant interviews. During focus group discussions, we also applied several **participatory rural appraisal tools (PRA)**, *such as perception mapping and historical timelines*, to facilitate discussions with villagers about their perceptions of natural resources and livelihood activities before and after displacement. Group discussions were also carried out with children to gather their attitudes about changes before and after displacement and issues related children, young people, and their families.

After focus group discussions, ***household surveys*** were conducted to quantify impacts of dam-induced displacement and resettlement on livelihood changes of displaced households and to determine opinions of households on decisions of land acquisition, compensation, and resettlement, livelihood supports, and benefit sharing. After household surveys, we selected several households and village and commune leaders or staffs to have ***in-depth interviews***. Furthermore, we also had many observations and informal talks with villagers to discover their personal views about impacts of resettlement on livelihoods. Additionally, information noted from conferences was very important to understand perspectives of different stakeholders who directly or indirectly engaged in hydropower development. More precisely, research tools and activities were described and explained for each chapter.

Data analysis: The study employed SPSS software to analyse quantitative data collected at household level (mainly for two case studies in chapters 4 and 5). To quantify the changes of displaced people's living conditions, before and after method was used to estimate changes. Surveyed households were categorized into different groups in terms of gender, age, social position, health status, and ethnicity of household heads. Descriptive analysis was used to display the socio-economic conditions of surveyed households before and after displacement. To compare among these groups, a paired sample T-test was used to determine mean income differences. Pearson correlation was tested to determine the relationship between income and other influential variables on livelihood change. Qualitative data was also coded to SPSS to estimate the descriptions of respondents. Both quantitative and qualitative data were converted to tables, graphs, and figures to interpret meaning and to answer pre-formulated research questions.

1.4 Structure of the book

The dissertation is structured as following chapters:

Chapter 1: Introduction. The chapter gives reasons for this study and provides the background of global-level hydropower debate. Research aims and questions are also formulated. The conceptual framework and study approaches are employed to lay a foundation for further analysis in succeeding chapters. Next, the reasons why Vietnam was selected for this study are explained. Finally, research methods and activities are generally defined because they are also explained for each chapter in specific.

Chapter 2: The evolution of hydropower and its emerging issues in Vietnam. Chapter 2 addresses the first sub-question. In this chapter, the benefits and costs of hydropower dam construction are described to identify emerging issues over hydropower development in Vietnam.

Chapter 3: Vietnam in the debate on land grabbing: conversion of agricultural land for hydropower development. Chapter 3 deals with the second sub-question, in which the first section presents an overview of agricultural land acquisition and conversion in recent years. The second section focuses on explaining why agricultural land has been converted to different purposes in Vietnam. The third section concentrates on the case of land conversion for hydropower development, in which it looks at the history of land conversion policy at different periods, describing the current land conversion mechanism for hydropower dam construction, and identifying shortcomings and consequences due to the conversion of agricultural land to hydropower dams. The last section concludes and discusses the material presented.

Chapter 4: Compensation and resettlement policies after compulsory land acquisition for hydropower development in Vietnam: policy and practice. Chapter 4 also addresses the second sub-question, but it focuses on a case study of a recently constructed hydropower dam in Central Vietnam, which followed the most recent policies on compensation and resettlement, and, therefore, was appropriate to provide an understanding of the recent changes in policy on hydropower development, land acquisition, compensation, displacement and resettlement, livelihood support, and benefit sharing. In this chapter, we also review important theories and concepts related to the issues of the compensation and resettlement. After that, we identified the gap between policy and praxis by presenting a case study of A Luoi dam construction. Finally, the root causes of problems are presented by conveying opinions from a former Vice-Minister of Ministry of Environment and Natural Resources (MONRE).

Chapter 5: Dam-induced displacement in Central Vietnam: vulnerability, inequality, and livelihood pathways after resettlement to the vicinity of urban fringe. Chapter 5 deals with the third sub-question on impacts of hydropower dam construction on displaced people's livelihoods and living conditions after resettlement. This chapter also aims to seek alternatives for better resettlement and more appropriate livelihood supports for displaced people due to hydropower dam construction, in particular, attention is paid to putting displaced people in new resettlement sites that they provide more opportunities to diversify their livelihoods rather than traditional resettlement programs that focus on restoring their farming activities only.

Chapter 6: The roles and influences of NGOs in governing hydropower development in Vietnam. Chapter 6 seeks answers for the fourth sub-question on the involvement and influences of NGOs on decision-making processes in hydropower development in Vietnam. Firstly, the analytical framework presents an argument regarding the relation between the state and civil society organizations

(CSOs) and NGOs under the single party-state countries. This argument lays theoretical foundation for analysing the NGOs-state relation in Vietnam. After that, the management system of NGOs in Vietnam is presented to understand more deeply the relations between NGOs and the state. More importantly, actions and activities of NGOs on the ground are presented in three case studies to examine their roles and influences in addressing hydropower development problems. Finally, the discussion and conclusion will anchor the main findings and draw lessons from the involvement of NGOs in hydropower development.

Chapter 7: Conclusion. Chapter 7 aims to answer the central questions by reflecting on all research findings of previous chapters. All elements of good governance are revisited to evaluate to what extent good governance has been taken into account in the policy-making and practical implementation of hydropower development in Vietnam. Future implications of better governance in Vietnam will be discussed. The lessons learnt from hydropower development governance of Vietnam may help authors make arguments and reflections on the global governance of hydropower development and provide lessons learnt for other developing countries.

References

Ansar, A., Flyvbjerg, B., Budzier, A., & Lunn, D. (2012). Should We Build More Large Dams? The Actual Costs of Mega-dam Development.

ADB - Asian Development Bank. (2004). *Governance: sound development management.* Asian Development Bank.

Bakis, R. (2007). The current status and future opportunities of hydroelectricity. *Energy Sources, Part B, 2*(3), 259-266.

Beckerman, W. (1995). How would you like your 'sustainability', sir? Weak or strong? A reply to my critics. *Environmental Values, 4*(2), 167-179.

Brown, A., Müller, S., & Dobrotkova, Z. (2011). Renewable energy: Markets And prospects by technology. *IEA Information Paper.*

Brundtland, G. H. (1987). Our Common Future, World Commission on Environment and Development (WCED).

Carney, D. (2003). *Sustainable livelihoods approaches: progress and possibilities for change.* London: Department for International Development.

Cernea, M. (1997). The risks and reconstruction model for resetting displaced populations. *World development, 25*(10), 1569-1587

Cernea, M. M. (2000). Risks, safeguards and reconstruction: a model for population displacement and resettlement. *Economic and Political Weekly,* 3659-3678.

Cernea, M. M., & McDowell, C. (Eds.). (2000). *Risks and reconstruction: Experiences of resettlers and refugees*. World Bank Publications.

Change, C. (2007). Synthesis Report: Contribution of Working Groups I, II and III to the Fourth Assessment Report of the Intergovernmental Panel on Climate Change Core Writing Team.

Communist Party of Vietnam, (2006). *Documents of the 10th Party Congress*. Ha Noi, Vietnam: Political publisher.

De Haan, L. J. (2000). Globalization, localization and sustainable livelihood. *Sociologia Ruralis, 40*(3), 339-365.

De Haan, L., & Zoomers, A. (2005). Exploring the frontier of livelihoods research. *Development and change, 36*(1), 27-47.

Declaration, M. (1989). The Manila Declaration on People's Participation and Sustainable Development.

Development Assistance Committee. (1999). OECD. DAC Guidelines for Gender Equality and Women's Empowerment in Development Co-Operation.

Edenhofer, O., Pichs-Madruga, R., Sokona, Y., Seyboth, K., Kadner, S., Zwickel, T., ... & Matschoss, P. (Eds.). (2011). *Renewable energy sources and climate change mitigation: Special report of the intergovernmental panel on climate change*. Cambridge University Press. (IPCC, 2011)

Ellis, F. (2000). The determinants of rural livelihood diversification in developing countries. *Journal of Agricultural Economics, 51*(2), 289-302.

EREN 2014 - Department of Energy, Energy Efficiency and Renewable Network (EREN) http://www.eren.doe.gov/consumerinfo/refbriefs/tphydro.html

EVN – Vietnam Electricity Group. (2009). *Vietnam Electricity – growth with the development of the country 2009*. Ha Noi, Vietnam: EVN

Flint, R. W., & Houser, W. L. (2001). *Living a sustainable lifestyle for our children's children*. IUniverse.

Gagnon, L., Klimpt, J. É., & Seelos, K. (2002). Comparing recommendations from the World Commission on Dams and the IEA initiative on hydropower. *Energy policy, 30*(14), 1299-1304.

Government of Vietnam. (2012). *Implementation of sustainable development strategies in Vietnam - National report at the United Nations Conference on Sustainable Development (rio+20)*. Ha Noi: Vietnam Government.

Hennink, M., Hutter, I., & Bailey, A. (2011). *Qualitative research methods*. Sage.

ICOLD. (2010). International Commissionon Large Dams. Retrieved from http://www.icold-cigb.org

IEA. (1990). *Energy balances of OECD countries 1987–1988*. International Energy Agency/Organisation for Economic Co-operation and Development, Paris

IEA. (2000). *Hydropower and the environment: present context and guidelines for future action.* IEA Hydropower Agreement Annex III.

IEA. (2008). *World Energy Outlook 2008.* Paris: OECD

IEA. (2010). Energy Technology Perspectives 2010: Scenarios & Strategies to 2050.

IEA. (2012). *Technology Roadmap – Hydropower.* Paris: OECD

IPCC - Intergovernmental Panel on Climate Change. (2007). *Climate Change 2007-Mitigation of Climate Change: Working Group III Contribution to the Fourth Assessment Report of the IPCC.* Cambridge University Press.

International Renewable Energy Agency (IRENA). (2012). Renewable Cost Database. Bonn: IRENA.

International River Network. (2014). *Southeast Asia Partner Organizations.* Retrieved from http://www.internationalrivers.org/resources/southeast-asia-partner-organizations-3598

IRENA, 2012. Renewable Energy Technologies: Cost Analysis series: Hydropower. http://www.irena.org/DocumentDownloads/Publications/RE_Technologies_Cost_Analysis-HYDROPOWER.pdf

Jabareen, Y. (2008). A new conceptual framework for sustainable development. *Environment, development and sustainability, 10*(2), 179-192.

Japanese Society of Energy and Resources, 1996. Handbook of Energy and Resources (Japanese), Ohmsha Co. Tokyo, Japan.

Jia, J., Punys, P., & Ma, J. (2012). Hydropower. In *Handbook of Climate Change Mitigation* (pp. 1355-1401). Springer US.

Kardos, M. (2012). The reflection of good governance in sustainable development strategies. *Procedia-Social and Behavioral Sciences, 58*, 1166-1173.

Kirdar, U. (1992). Change: Threat or opportunity for human progress? V. 5. Ecological change: Environment, development and poverty linkages.

Koch, F. H. (2002). Hydropower—the politics of water and energy: introduction and overview. *Energy Policy, 30*(14), 1207-1213.

Korten, D. C., & Klauss, R. (Eds.). (1984). *People-centered development: contributions toward theory and planning frameworks.* West Hartford, CT: Kumarian Press.

Korten, D. C. (1990). Getting to the 21st Century. W Hartford, CT: Kumarian Press. pp. 3–4, 67–71.

Leach, M., Mearns, R., & Scoones, I. (1997). *Environmental entitlements: a framework for understanding the institutional dynamics of environmental change.* Brighton: Institute of Development Studies, University of Sussex.

McCully, P. (1996). *Silenced rivers: the ecology and politics of large dams.* Zed Books.

McDowell, C. (2002). Involuntary resettlement, impoverishment risks, and sustainable livelihoods. *The Australasian Journal of Disaster and Trauma Studies, 2*, 2002-2.

Mitigation, C. C. (2011). IPCC special report on renewable energy sources and climate change mitigation.

National Assembly of Vietnam. 2013. *The Constitutional of Vietnam 2013*. Ha Noi: National Assembly of Vietnam

Neumayer, E. (2003). *Weak versus strong sustainability: exploring the limits of two opposing paradigms*. Edward Elgar Publishing.

Nüsser, M. (2003). Political ecology of large dams: a critical review.*Petermanns Geographische Mitteilungen, 147*(1), 20-27.

Nwanze, K. F., & Kouka, P. J. (2010). International fund for agricultural development. In *Proceedings of the 3rd International Rice Congress*.

OECD Development Assistance Committee. (1996). *Shaping the 21st Century: The contribution of development co-operation*. 1996.

Ribot, J. C., & Peluso, N. L. (2003). A Theory of Access*. *Rural sociology, 68*(2), 153-181.

Schumann, K., Lau Saili, R. T., & Abdel-Malek, R. Hydropower and Sustainable Development: A Journey. Retrieved from http://www.worldenergy.org/documents/congresspapers /392.pdf

Scoones, I. (1998). Sustainable rural livelihoods: a framework for analysis.

Scudder, T. (2005). *The future of large dams: Dealing with social, environmental, institutional and political costs*. Earthscan.

Singh, S. (1997). *Taming the waters: The political economy of large dams in India*. Delhi: Oxford University Press.

Sovacool, B. K., & Bulan, L. C. (2011). Behind an ambitious megaproject in Asia: The history and implications of the Bakun hydroelectric dam in Borneo. *Energy Policy, 39*(9), 4842-4859.

Stern, N. (2008). The economics of climate change. *The American Economic Review*, 1-37.

Stone, R. (2010). Along with power, questions flow at Laos's New Dam. *Science, 328*(5977), 414-415.

Stone, R. (2011). The Legacy of the Three Gorges Dam. *Science, 333*(6044), 817-817

Termeer, C. J. A. M., Dewulf, A., & Van Lieshout, M. (2010). Disentangling scale approaches in governance research: comparing monocentric, multilevel, and adaptive governance. Ecology and Society, 15(4), 29.

The General Assembly of United Nations. (September 8, 2000). United Nations Millennium Declaration. United Nations. Retrieved from http://www.un.org/millennium/declaration/ares552e.htm

UNDP. (1997). *Governance for Sustainable Human Development*. UNDP Policy Paper.

UNDP. (2000). *Human Development Report: Human rights and human development. Summary. 2000*. Oxford University Press.

UNDP. (2011). *People-centred development: Empowered lives. Resilient nations.* New York: Communications/Partnerships Bureau United Nations Development Programme

Vandal, T. (2012). Hydroelectricity: green and renewable. *POLICY, 1.*

Wang, P., Dong, S., & Lassoie, J. P. (2014). *The Large Dam Dilemma: An Exploration of the Impacts of Hydro Projects on People and the Environment in China.* Springer.

WCED - World Commission on Environment and Development. (1987). *Our common future* (Vol. 383). Oxford: Oxford University Press.

World Bank (2009). Directions in Hydropower: Scaling Up for Hydropower. Water Sector Board Practitioner Notes (P-Notes), Sustainable Development Network of the World Bank Group, Washington, DC, USA, 16 pp.

World Bank. (1994). *Governance: the World Bank's experience.* World Bank.

WCD - World Commission on Dams. (2000). *Dams and development: a new framework for decision-making.* London London: Earthscan.

World Water Council. (2003). *Final Report of the 3d World Water Forum.* Kyoto, Japan: WWC. Retrieved from http://www.worldwatercouncil.org/fileadmin/world_water_council/documents/world_water_forum_3/3d_World_Water_Forum_Final Report_BD.pdf

2   THE EVOLUTION OF HYDROELECTRICITY IN VIETNAM AND EMERGING ISSUES[6]

2.1   Introduction

For most, electricity is an indispensable resource. Between 1980 and 2012, net worldwide electricity consumption[7] increased 62% (USEIA, 2012), and it is expected to grow 2.5% annually to 2030 (IEA, 2013). In 2010, electricity generated from fossil fuels (coal, gas, and oil) dominated the global electricity supply, accounting for 67.5% of total supply. Nuclear-based electricity supplied 12.9% of worldwide electricity, while renewable energy sources produced nearly 20%, most of which consisted of hydropower at 16% of global electricity production in 2010, or 3,423TWh (IEA, 2012). If the average wholesale price of electricity was estimated at US$19 cents per kWh for 35 countries between 2007 and 2012 (Nang Luong Viet Nam, 2013), then the 2012 worldwide electricity production has an approximated a value of US$650 billion, slightly more than the GDP of Switzerland in the same year (United Nation, 2013). Taking the lower wholesale electricity price of Vietnam as the point of departure (US$7.3 cents per kWh in 2012) (Nang Luong Viet Nam, 2013), then global electricity production corresponded to US$250 billion, which was 1.6 times higher than Vietnam's GDP (US$155 billion in 2012) (United Nation, 2013). On a national scale, hydroelectricity is particularly important for economic performance in 65 countries that rely on over 50% of their electricity supply from hydropower, while 32 countries depend for more than 80% (IHA, 2014). Globally, only 23% of the theoretical hydropower potential is currently exploited, which means that, in principle, some 5,400TWh/yr could be added or about twice the currently installed capacity (IHA, 2000). This could, theoretically, increase revenue by more than US$1026 billion per year at US$19 cents per kWh. If exploited properly, hydropower could help deliver electricity to 1.2 billion people currently living without electricity and, at the same time, enable access to water resources for 1.1 billion people (IEA, 2013). Particularly, some 65% of Africans lack access to electricity and, consequently, tend to live in poor conditions with poor access to lighting, clean water, health care, and education (ICOLD, 2008).

Vietnam has become an important global hydropower producer. Since independence in 1945, Vietnam has experienced nearly 70 years of hydropower

---

[6] *This chapter will be submitted as* Ty, P. H., Zoomers, A., & Van Westen, A. C. M. (2014). The evolution of hydropower in Vietnam and emerging issues. *Renewable and sustainable energy reviews*
[7] For all other countries except the United States, total electric power consumption = total net electricity generation + electricity imports – electricity exports – electricity transmission and distribution losses. http://www.eia.gov/cfapps/ipdbproject/docs/IPMNotes.html#e2

development and is quickly adding to its generating capacity. In 2011 and 2012, Vietnam contributed more to the global increase in hydroelectricity production (8% of worldwide growth, equivalent to 2GW in 2011, and 6% or 1.8GW in 2012) than any other country except China. This significant contribution was generated by the Son La hydropower plant, the largest hydropower dam in Southeast Asia, completed and put into operation in 2011, with an installed capacity of 4000 MW (REN21, 2012).

In Vietnam, hydropower accounts for over 40% of total electricity production, which ranks second after fossil fuel generated electricity (CODE, 2010). In the electricity development plan for the period up to 2020, hydropower production is set to increase from 9200 MW in 2010 to 17400 MW[8]. Production at this level calls for a review of hydropower development in Vietnam, appraising its significance and dynamics and identifying issues that need to be considered in view of future development. The data and information for this paper was collected from metadata sources, literature, conference notes, online newspapers, and government reports.

The paper is organized as follows. Firstly, the evolution of hydropower and its drivers are identified. Secondly, the contribution of hydropower to the socio-economic development is analysed. Thirdly, environmental, social, and political issues are discussed. Finally, lessons learnt from experiences are drawn and policy implications proposed.

2.2 The drivers and evolution of hydropower in Vietnam

Vietnam's early experiences with electricity are reflected in the story *'upside down lamp – cây đèn lộn ngược'*, which recounts officials of the Nguyen Dynasty who first saw electric streetlights when visiting France in 1863 In 1892, the French built the first small electricity generating plant (nhà máy đèn) in Ha Noi to serve 523 streetlights. In 1943, the first hydropower dam, Ankroet in Lam Dong province, was constructed with an installed capacity of 2.3MW, which mainly served schools, hospitals, and especially French hotels in the city of Da Lat (Lam Dong PPC, 2014). After independence in 1945, the state placed electricity generation high on the agenda of public investment: "electricity must go earlier than other policies." For rural development, the phrase *"điện, đường, trường, trạm - electricity, roads, schools, stations (clinic, post office, and others)"*has become the slogan for public investment in rural areas. In national electricity development planning since 1945, hydropower is often emphasized, and it has evolved along with the *Doi moi*—economic reform process.

---

[8] Approval decision of the national master plan for power development for the 2011-2020 period with the vision to 2030

Between 1945 and 1975, the period of the Vietnam War, only two hydropower dams were constructed. The first one, Da Nhim, was put into operation in 1964with Japanese governmental funding south of the Dong Nai river basin, with a total installed capacity of 160MW, which generates over 1000GWh/year. The Da Nhim reservoir supplies more than 550 million cubic meters of water each year for irrigation of 20000 ha of arable land in Ninh Thuan Province, the province with the lowest average annual rainfall. The second hydropower plant, Thac Ba in the north, was completed in 1975 after 12 years of construction with an installed capacity of 120MW, which that generated 70% of total electricity production throughout the North at the time. It was devastated repeatedly by war, but restored with great pains under the slogan *"The nation needs electricity as the body needs blood"* (Nguyen Hoai Nam, 2010).After reunification in 1975, several large hydropower projects, such as Hoa Binh (1979) and Tri An (1984), were launched but not completed in the period up to 1986, as this was the period of least dynamic development in Vietnam.

Between 1986 and 1994, the first stage of economic reform took place in Vietnam. In this period, the Hoa Binh hydropower plant was put into operation in 1994 after 15 years of construction, which, at the time, was the largest hydropower dam in Southeast Asia with an installed capacity of 1,920MW. To complete this dam, Vietnam mobilized a huge amount of labour and money. Construction displaced 9,305 households with more than 58,000 people. Twenty-four communes were flooded and 13,000 hectares of land lost. Fifty thousand workers were mobilized for the project, and Hoa Binh province provided about 8,000 tons of rice, 2,650 tons of pork, thousands of tons of vegetables for 50,000 workers who worked day and night. Many people were excited and voluntarily dedicated their efforts to the Hoa Binh dam (Hoang Viet Cuong, 2009).This shows that most people accepted the project, even though they had to sacrifice in terms of living standards in the hope of a better future for the country and themselves.

Once in operation, the Hoa Binh hydroelectric plant contributed about 9 GWh per year, which accounted for 30-40% of overall electricity production. The contribution of the Hoa Binh hydropower plant together with thermal power plants exceeded the power demand in the north. Therefore, a north-south, high-voltage grid was built with a length of 1,487km to provide electricity to the central and southern regions and it completed in1994 at the same time as the Hoa Binh dam (Hoang Viet Cuong, 2009). Together with the Tri An hydropower project, the second largest dam completed in 1991 with an installed capacity of 400MW and has an annual electricity output of 1700GWh supplying Ho Chi Minh city (Tu, et al., 2011). Hoa Binh stood out as a symbol for the era of accelerating growth and unchallenged modernization in Vietnam in the 1990s, the 'happy days' when the renovation process ushered in a new era of progress and pride (EVN, 2009).

Between 1995 and 2005, economic growth increased to an average of 7.5% per year, whereas the demand for electricity expanded at 15% per year. Therefore, Vietnam embarked on large hydropower dam projects in the Central Highlands and the Dong Nai river basin in the South, including Yali (180MW),[9] Vinh Son (66MW), Song Hinh (70MW), Thac Mo (150MW), Ham Thuan (300MW), and Da Mi (175MW) (Institute of Energy, 2006). The state-owned companies constructed the majority of these dams. However, there have been many important changes in the investment policies of hydropower development along with achievements in social and economic development and international integration of Vietnam in the following period.

From 2006 to present, Vietnam set an even more ambitious electricity development plan. The government prepared necessary policies, legal instruments, and human forces for a new building plan; for example, Electricity Law (2004); Land Law (2003) and its lower guidelines; Environmental Protection Law (2005); Law on Competition (2006); Business Freedom policy for members of Communist Party (2006); WTO membership (2006); and other policies. As a result, Electricity Plan 6 for the period from 2006-2011 and Plan 7 for the period between 2011 and 2015 toward 2030 was formulated to promote electricity development in which hydropower was prioritized. However, the government did not have sufficient financial resources for implementation, and, therefore, the government opened the process to allow private sector investment in the construction of small and medium-sized hydropower dams. This strategy was consistent with the liberalization policy of the competitive electricity market. At the same time, the government decentralized investment licensing, project approval, and hydropower development planning to provincial governments. The resonance of these conditions became the catalyst for the booming of hydropower projects across the country to meet the increases of electricity demand of 13.8% per year (Khanh Toan, et al., 2011). As a result, 32 large hydropower dams and 16 medium hydropower plants were constructed in this period. Particularly, the largest Son La hydropower dam with an installed capacity of 2400MW was put into operation in 2012, considered as the largest hydropower dam in the South East Asia.

Figure 2-1 shows that hydropower dams are constructed mainly in the Northern and Central highland regions of Vietnam. According to the survey by the Ministry of Industry and Trade (MOIT), 260 hydropower plants currently operate with an installed capacity of 13,694 MW, 211 projects are under construction and will be in operation by 2017 with an installed capacity of 6,712MW, and 266 future projects with an installed capacity of 3,410MW are being considered to license to investors (MOIT, 2013). In 2013, total hydroelectricity production of Vietnam was 56.94TWh, representing nearly 70% of Vietnam's theoretical hydropower potential

---

[9] The installed electricity capacity of power stations

(see Table 2-1). By 2017, Vietnam will have exploited nearly all its theoretical hydropower potential, compared to a global average of 35% of exploited hydropower potential (IHA, 2014).

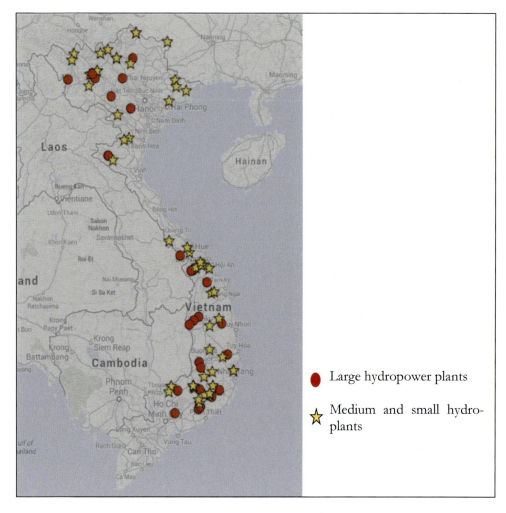

Figure 2-1 Location of operating hydropower dams in Vietnam.

Source: Authors' research from the internet

Table 2-1 Hydroelectricity potentials in main river watersheds in Vietnam.
Source: EVN, 2009

	River basin	Region	Potential installed capacity (MW)	Theoretical hydroelectricity potential (TWh)	(%)
1	Lô – Gấm – Chảy	North	1,120	4.10	4.9
2	Đà	North	6,960	26.96	32.3
3	Mã	North	890	3.37	4.0
4	Cả	North	520	2.09	2.5
5	Vu Gia – Thu Bồn	Central	1,360	5.10	6.1
6	Trà Khúc - Hương	Central	480	2.13	2.6
7	Ba	Central – Highland	670	2.70	3.2
8	Sê san	Central – Highland	1,980	9.36	11.2
9	Srêpok	Central – Highland	700	3.32	4.0
10	Đồng Nai	South and Central Highland	2870	11.64	14.0
Ten watersheds			**17,550**	**70.77**	**84.5**
All country			**20,560**	**83.42**	**100**

To 2030, Vietnam will continue to place priority on building hydropower dams, focusing on small hydro projects. Currently, 179 small projects (2360MW) are under construction, 249 projects (2327MW) are to be approved, and there are available155 potential locations for further development. Vietnam aims to exploit completely the theoretical hydroelectricity potential to meet a forecasted demand of electricity consumption, which is expected to increase 1.6 times by 2020 and 4.5 times by 2030, as compared to current electricity demand.

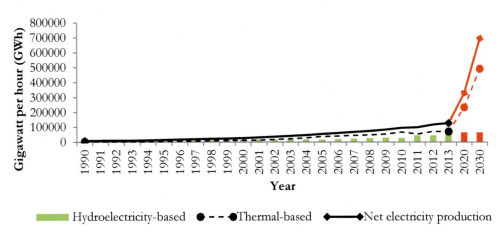

Figure 2-2 Electricity production from 1990 to 2013 and forecasted (in red lines) to 2030. *Source: EVN, 2009; United Nation, 2013*

As Figure 2-2 shows, the role of hydroelectricity will remain important until 2030. However, after 2020, the national electricity grid will depend on a mostly fossil-fuel source-based electricity production. Therefore, Vietnam has begun to build hydropower dams in Laos and Cambodia in 2010 to supplement the national electricity grid (MOIT, 2013). Thus, the hydropower development plan of Vietnam is almost complete. With these considerations, a comprehensive assessment is necessary to detect the benefits and consequences of hydroelectric development.

2.3 Benefits of hydroelectricity development in Vietnam

The function of hydropower dam construction is to ensure electricity security, which, in turn, promotes national socio-economic development. Therefore, this paper provides an examination of the relationship between electricity security and its influences on social and economic development. Firstly, electricity security was officially affirmed by the general manager of Vietnam Electricity Group (EVN) in 2013, when he declared that EVN could supply sufficient electricity for national socio-economic development, with a reserve at 20% of total electricity demand (Bich Diep, 2014; MOIT, 2013). Total commercial power in 2013 was 115 billion kWh, which was equivalent to US$ 8.7 billion (at 7.5 cents per kWh), corresponding to 5% of GDP in 2013. In particular, hydropower occupied over 66% of total annual power production between 1990 and 2002 and nearly 37% from 2003 to date. From 1990, the average annual hydroelectricity production increased by 10%, which is greater than 7% of the average annual GDP for the same period.

Besides the economic value, the strength of the growth of electricity and hydroelectric has enabled Vietnam to strengthen an already strong capacity for the electricity sector. Currently, EVN employs more than 97,000 employees with high salaries compared to other industries (EVN, 2009). Additionally, the Son La hydroelectric plant employed 12,000 workers and engineers for dam construction between 2005 and 2012. The skills of Vietnamese constructors have greatly improved, so that they are able to complete large-scale hydropower projects independent of international expertise. For example, the largest Son La hydropower plant was consulted by the Vietnamese Power Engineering Consulting Joint Stock Company 1 (PECC1) and constructed by the Song Da Corporation and other Vietnamese sub-contractors. All tasks—planning, designing, construction, and operation—were used to be done by foreign companies, but now mostly carried out by Vietnamese companies.

Moreover, electricity growth has contributed to a speedup of industrialization and modernization processes in Vietnam. Figure 2-3 shows that electricity consumption for industry increased significantly. In 2012, industry and construction consumed more than 52% of total national electricity production and households consumed over 40%; the service sector also tended to use more electricity. The increase of electricity consumption has also increased the contribution of industry and service for the national economy. Figure 2-4 indicates that there was a considerable increase in the value of industry and services, whereas there was a significant decrease in the value of agriculture sector. This was an expected trend for the transition of Vietnam economy toward 2020 as an industrialized country.

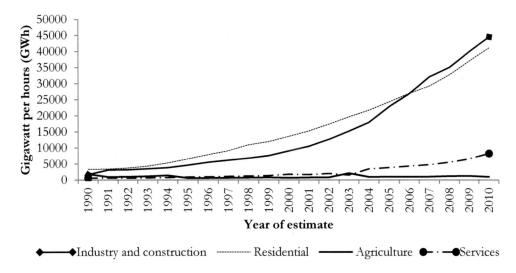

Figure 2-3 Electricity consumption of different sectors from 1990 to 2010.

Source: United Nation, 2013

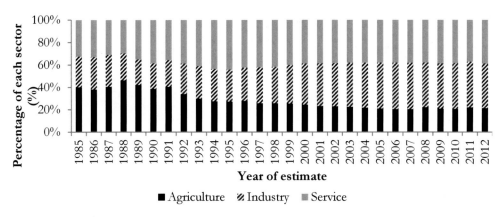

Figure 2-4 Values of industry, service, and agriculture sector.

Source: United Nation, 2013

31

In conjunction with the above, hydroelectricity has contributed considerably to economic growth (Table 2-2). It was assumed that hydropower has contributed nearly 50% of electricity production for the whole development period, so that it also has added 50% of effect to economic growth. Herein, we applied Pearson correlation to quantify the strength of the relationships between electricity and economic growth variables at a national scale. We also assumed a linear relationship between electricity and economic growth variables. The Pearson correlation analysis showed a strongly positive relationship between an increase of electricity, hydroelectricity production, and electricity consumption per capita with the growth of socio-economic development variables (See Table 2-2). It means that electricity production growth supports economic development. The analysis also indicated that electricity production growth is negatively correlated with poverty and population growth. It means that when electricity production increases, poverty and population growth rate decreases.

More importantly, the noticeable contribution of the electricity sector was the success of the rural electrification program. It was widely recognised by many international organizations, such as World Bank, ADB, as lessons for effective rural electrification programs for other developing countries (ADB, 2011; EVN, 2009; Minh Do and Sharma, 2011; MOIT, 2013). After reunification in 1975, only 2.5% of the rural population in Vietnam could access electricity, which increased to over 97% by 2012. In urban areas, nearly 100% of households had access to electricity (EVN, 2013). The electrification rate was higher than China and Southeast Asian countries, such as Thailand and Malaysia, in the same period, whereas Vietnam had a lower GDP per capita than these countries (ADB, 2011; World Bank, 2011).

Table 2-2 Correlation between electricity and economic growth variables

		GDP	GDP per capita	Poverty Rate	Population growth	Mobile use	Life expectancy	Water	Sanitation	FDI growth
Electricity consumption per capita	R	.974**	.992**	-.842**	-.809**	.915**	.684**	.950**	.955**	.892**
	Sig	.000	.000	.000	.000	.000	.000	.000	.000	.000
	N	28	43	21	42	38	41	22	22	26
Hydro-electricity growth	R	.972**	.967**	-.817**	-.832**	.888**	.937**	.967**	.969**	.826**
	Sig	.000	.000	.000	.000	.000	.000	.000	.000	.000
	N	22	22	20	22	22	21	21	21	21
Total electricity growth	R	.991**	.987**	-.844**	-.866**	.927**	.893**	.952**	.957**	.864**
	Sig	.000	.000	.000	.000	.000	.000	.000	.000	.000
	N	23	24	21	23	23	22	22	22	22

**Correlation is significant at the 0.01 level (2-tailed); R- Pearson correlation coefficient; Sig-significant (2-tailed); N -number of estimated years

In recent years, rural electrification completely changed the face of rural Vietnam, improving quality of life and enhancing crop yields and livestock levels by providing sufficient water, as well as improving infrastructure, mechanization, irrigation, and agro-processing products, while enabling new livelihoods, such as ice cream making and sewing. Rural electrification has improved the quality of medical services, education, and drinking water supply, as well as increasing access to market information, allowing for a more effective means to disseminate agricultural knowledge and public policy, increasing incomes, and reducing costs of production. For example, 92% of rural households in Gia Lai and Thai Nguyen province confirmed that the introduction of electricity increased the incomes of those using tea processing machines. Income growth was based on an increase in processing efficiency and also renting the tea processing machines to others (Kooijman-van Dijk & Clancy, 2010). Levels of education of adults and children in electrified households were higher, as compared to unelectrified households (World Bank, 2011). Further, adequate electricity supplies enabled rural people to stop logging woods for cooking, which, in turn, could help forest and environmental protection (Kooijman-van Dijk & Clancy, 2010).

Beyond electricity generation, the majority of hydropower reservoirs contributed considerably to multiple purposes, such as flood and drought control, irrigation, and infrastructure and tourism development. For example, Son La hydropower plant not only generated over 10 billion kWh of electricity but also contributed to flood control for the entire northern plains with a storage capacity of 7 billion m³. This reservoir irrigated tens of thousands of hectares of rice. Further, Son La hydropower built more than 125 km of rural roads, many bridges, and telecommunication systems throughout the region (EVN, 2009; MOIT, 2013). Capital investment for resettlement and infrastructure accounted for 31% of total construction cost, about US$570 million—a large investment.

Above are advantages of hydropower development. Below is an examination of disadvantages, which will help in understanding why hydropower development is a controversial issue in Vietnam.

2.4   Emerging issues

Behind the significant contribution of hydropower in Vietnam, in recent years hydropower development has caused many sever adverse environmental and societal consequences, which have become controversial issues in political forums, newspapers, and scientific conferences, from national to local levels (Bui, et al., 2013; CODE, 2010; Dao, 2011; EVN, 2009; Minh Do & Sharma, 2011; Ty, et al., 2013).

2.4.1 Dam failures and water conflicts

In Vietnam, large hydropower dams are often carefully surveyed, consulted, designed, constructed, and monitored by highly competent experts and contractors from qualified hydropower development countries, such as Sweden, United States, Japan, China, Switzerland, and Russia. To date, failures of large dams have been effectively avoided (Pham Hong Giang, 2013).

However, the development of small and medium-scale hydropower dams has caused problems since the central government gave the right to provincial authority for granting investment license to private companies to build small and medium-scale hydropower dams in their respective administration boundary in 2005. Since then, the provinces have constructed many small and medium hydropower dams. According to Ministry of Industry and Trade (MOIT), Vietnam has 2005 small hydropower plants under operation with a capacity of 1,664MW, and 179 projects are under construction with a capacity of 2,360 MW to be in operation by 2016. Additionally, 249 projects are being considered for investment licenses, and an additional 155 potential locations are being planned in several provinces (MOIT, 2013).

Most small hydropower projects did not strictly comply with dam safety measures regulated by provision of laws, leading to unpredictable risks and failures. The report of the Committee on Science, Technology and Environment of the National Assembly revealed that nearly 30% of small hydropower dams were not technically verified. Therefore, many small dams have failed causing substantial damages to property, affecting people's lives in downstream regions. For example, in 2012 in Kon Tum province, the dam break of Dak Me 3 caused one death and several injuries (Huu Kha and Le Trung, 2012). Also in 2012, the break of Dakrong 3 damaged a large area of cassava land in Quang Tri (Tran Tinh, 2012). The failure of Ian Keel 2 dam destroyed 200 hectares of cropland (Thien Thu, 2013). These failures were mainly caused by poor preparation and construction by investors and poor management by local authorities (Le Anh Tuan & Lam Thi Thu Suu, 2013; Pham Hong Giang, 2013). In addition, for small and medium dams, approximately66% lacked safety plans and 55% lacked flood prevention plans (Hoang Linh, 2014). Most of medium and small hydropower plants were considered as multi-purpose dams in the planning stages serving not only for electricity generation but also as irrigation sources and flood control, but most investors follow through on these plans (Pham Hong Giang, 2013). For example, Ba Ha hydropower plant decreased the flood storage of reservoir from 256 million m^3 during planning stage to 167 million m^3, and they continued reducing flood capacity to a very low level (Institute of Energy, 2011).As a consequence, small and medium hydropower plants must discharge water during flooding period leading to downstream flooding. Conversely, dams often store water in summer season for electricity generation leading to downstream drought and salinity intrusion (Hoang Linh, 2014; Institute of Energy, 2011; Pham Hong Giang, 2013). Recently, these situations have led to several water disputes between upstream and downstream provinces: for example, the water dispute between the city of Da Nang and the Dak Mi 4 hydropower project in Quang Nam province. This dispute was a result of Dak Mi 4 hydropower blocking water flow downstream to Da Nang city where more than 1.7 million people depended on this water for drinking and for agriculture production on 20,000 ha of land, as well as for navigation around the most important docks in Vietnam's central region (Institute of Energy, 2011). Another severe problem is earthquakes initiated during reservoirs closures to store water for large hydropower dams. Since 2011, Song Tranh 2 hydropower dam, for example, has caused many earthquakes in Quang Nam province, which disturbed the lives of people in this region (Le Ha, 2013).

2.4.2   Loss of land and valuable ecosystems

To date, over 620,000 ha of land has been lost, in which small hydropower dams accounting for 6% of this total, where forest and agricultural land accounts for the highest percentage of lost land usage. Meanwhile, hydropower investors neglected reforestation, and the loss of agricultural land to farmers was not adequately compensated after resettlement. This has caused long-lasting grievances and denouncements from local to national levels related to land acquisition for hydropower dams, inadequate compensation, and poor management of resettlement programs (CODE, 2010; Dao, 2010; Institute of Energy, 2011; MOIT, 2013; Ty, et al., 2013).

Recently, as the number of small hydropower dams has rapidly increased, many thought that small dams required little forested land, but, in fact, these dams destroyed large areas of forest. For example, Ha Nang hydropower dam with an installed capacity of 11MW caused 36.15 ha of natural forest loss; Dak Ru dam was installed with a capacity of 7.5MW but destroyed more than one hundred hectares of natural forest; and Dak Po Co cleared 117.2 ha of natural forested land. According the strategic environmental impact assessment of the electricity development plan from 2011 to 2015 and toward 2030, the total area of lost land will be approximately 25,133 ha flooded by 21 large hydropower dams, including 4,227 ha of natural forests, 1,367 ha of plantation forest, 5,961 ha of agricultural land, 737 ha of residential land, and 12,810 ha of pastures and bared land (Institute of Energy, 2011). The loss of land caused the substantial loss of rich biodiversity land areas as well (see Table 2-3). According to Decree 99/2010/ND-CP dated 24/09/2010 on environmental service payment (PES) for forest protection, hydropower projects must pay US$0.3 cent per kWh. The value of environmental service payments of hydropower was estimated over US$2 million in 2011, but hydropower owners haven't paid sufficiently (Institute of Energy, 2011).

Table 2-3 Affected land area of highly biodiversity natural reserves.
Source: Institute of Energy, 2011

Hydropower projects	Nature protected areas	
	Name	Affected areas (ha)
Bac Me	Bac Me	24,238
	Du Gia	1,144
Ban Chat	Nam Don	3,592
Dak Mi 1	Ngoc Linh	23,061
	Song Thanh	9,541
Dong Nai 2	Tay Nam Lam Dong	1,168
	Song Thanh	6,601
Dong Nai 5	Cat Tien	19,092
	Tay Nam Lam Dong	9,798
Hua Na	Mu Cang Chai	298
	Nam Don	21,144
Lai Chau	Muong Nhe	77,968
Srepok 4	Yok Don	20,229
Trung Son	Pu Hu	12,533
	Pu Luong	1,025
	Xuan Nha	11,298
Total affected of protected areas		**540,432**

2.4.3 Displacement and deterioration of resettled people's living standards

To 2013, construction of hydropower dams has displaced about 60,000 households and 240,000 people. From 2013 to 2030, a further 21 large hydropower projects will displace over 60,000 people. Thus, hydropower development will displace approximately 300,000 people (approximately 0.3% of the total population) of which 90% are poor ethnic minority groups relying on forest and agriculture livelihoods and face many challenges to sustain their lives after resettlement (Bui & Schreinemachers, 2013; Dao, 2010; Institute of Energy, 2011; Ty, et al., 2013). Research by the Institute of Development Consultation – (CODE, 2010) showed that over 82% of resettled people became worse off after resettlement. This finding is also similar to our recent survey in the A Luoi resettlement site in Thua Thien Hue province (Ty, et al., 2013).

The poverty rate of displaced communities was persistently higher than the rest of Vietnam (11 %) in 2009 (see more in Table 2-4). As a result of insufficient agricultural land (especially a lack of paddy fields), poverty is often accompanied by hunger, as most families cannot produce enough food (CODE, 2010; Ty, et al., 2013).

Table 2-4 Poverty rate of displaced households due to large hydropower dams. *Source: The management board of the project 747 and 472 (2009)*

Classification	Hydropower dam				Average rate (%)
	Hoa Binh	Ban Ve	Yaly	Tuyen Quang	
Poor	33.4	81.1	65.6	46.3	56.3
Near poor	20.3	3.5	0.7	20.6	11.5
Average	44.9	15.4	33.7	33.1	31.8
Better-off	1.4				0.4

2.5   Conclusion

The findings show that Vietnam has implemented a long-term program of hydropower development as a means to fuel industrialization and modernization. The development of hydropower in Vietnam has contributed significantly to the economic growth and structural changes. However, increasingly negative impacts on environment and society, especially displaced communities, have emerged to date. The case of hydropower development in Vietnam also changes our mind about small hydropower dams that they cause less problems to environment and society. In practice, small hydropower dams also cause a lot of adverse environmental and social impacts.

Increasingly adverse impacts of hydropower dams on environment and society have changed people's perception on hydropower. Hydropower, once a source of pride, is now a fear. This change also comes from the economic and social transformation when Vietnam's economy grows and living standards are improved, while environmental and social concerns are increasing. Moreover, with market economy reforms and decentralized governance, private sectors and local authorities often implement hydropower projects with less attention to social and environmental consequences. Many private companies became richer by exploiting natural resources at low costs, but this exploitation caused problems to environment and to vulnerable people without adequately redistributing benefits to those affected. Therefore, the involvement of private companies in hydropower investment has been criticized by civil society organizations, media and activists who are working to protect vulnerable people. Beyond these facts, discussion has now shifted from a top-down, modernization views of development focus on economic growth, technology, and machinery towards bottom-up, post-modernization views that puts more emphasis on environmental values, local communities, identities, and equity.

References

ADB. (2011). *Energy for all: Viet Nam's success in increasing access to energy through rural electrification.* Asian Development Bank: The Philippines.

Bich Diep. (2014). Tổng giám đốc EVN "trải lòng" về ghế nóng. 05/06, 2014, Retrieved from http://m.dantri.com.vn/kinh-doanh/tong-giam-doc-evn-trai-long-ve-ghe-nong-835151.htm

Bui, T. M. H., Schreinemachers, P., & Berger, T. (2013). Hydropower development in Vietnam: Involuntary resettlement and factors enabling rehabilitation. *Land use Policy, 31*(0), 536-544. doi:10.1016/j.landusepol.2012.08.015

CODE. (2010). *Displacement, resettlement, living stability and environmental and resources protection in hydropower dam projects.* CODE: Ha Noi, Vietnam.

CODE, The center for development consultation. (2010). *Study the situation and effects of thuong kon tum on socio-economic and environmental factors - the proposal of displacement and resettlement consultation for the thuong kon tum hydropower dam.* CODE: Ha Noi.

Dao Trong Tu, Le Thi Thuy Quynh, Pham Quang Tu, and Bach Tan Sinh. (2011). *Sustainability assessment of Vietnam's electricity planning – using section 1 of the 2009 hydropower sustainability assessment protocol.* . Ha Noi, Vietnam.

Dao, N. (2010). Dam development in Vietnam: The evolution of dam-induced resettlement policy. *Water Alternatives,* 3(2), 324-340.

Dao, N. (2011). Damming rivers in Vietnam: A lesson learned in the tay bac region. *Journal of Vietnamese Studies,* 6(2), 106-140. Retrieved from http://search.proquest.com/docview/902084590?accountid=14772

EVN. (2009). *Vietnam electricity – growth with the development of the country.* EVN: Ha Noi, Vietnam.

EVN. (2013). Thủy điện sơn la: Về đích trước 3 năm, làm lợi 1 tỷ usd. 05/20, 2014, Retrieved from http://pcphuyen.cpc.vn/?show=news&catid=11&contentid=1688

Hoang Linh. (2014). Small hydropower dams, big issues. *Journal of New Energy, 311*

Hoang Viet Cuong. (2009). Hoa binh hydropower – the great work of 20th century. 06/10, 2014, Retrieved from http://soxaydung.hoabinh.gov.vn

Huu Kha and Le Trung. (2012). Xe ben đụng vỡ... đập thủy điện. 05/18, 2014, Retrieved from http://tuoitre.vn/Chinh-tri-Xa-hoi/522134/Xe-ben-dung-vo-dap-thuy-dien.html#ad-image-0

ICOLD. (2008). *World declaration: Dams and hydropower for African sustainable development*

IEA. (2012). *World energy outlook 2012*. International Energy Agency (IEA): Wellington, New Zealand.

IEA. (2013). *Key world energy statistics*. International Energy Agency: France.

IHA. (2000). *Hydropower and the world's energy future*. International Hydropower Association (IHA): United Kingdom.

IHA. (2014). *Advancing sustainable hydropower*. International Hydropower Association (IHA): United Kingdom.

Institute of development - CODE, Vietnam. (2010). *Displacement, resettlement, living stability and environmental & resources protection in hydropower dam projects*. Institute of development - CODE, Vietnam: Ha Noi.

Institute of Energy. (2006). *The master plan on power development for the period of 2006–2015 with perspective to 2025*. Hanoi, Vietnam.

Institute of Energy. (2011). *Strategic environmental assessment for the electricity plan between 2011 and 2015 and toward 2030*. Energy Institute: Ha Noi, Vietnam.

Khanh Toan, P., Minh Bao, N., & Ha Dieu, N. (2011). Energy supply, demand, and policy in Viet Nam, with future projections. *Energy Policy, 39*(11), 6814-6826. doi:http://dx.doi.org/10.1016/j.enpol.2010.03.021

Kooijman-van Dijk, A. L., & Clancy, J. (2010). Impacts of electricity access to rural enterprises in Bolivia, Tanzania and Vietnam. *Energy for Sustainable Development, 14*(1), 14-21. doi:http://dx.doi.org/10.1016/j.esd.2009.12.004

Lam Dong PPC. (2014). Power. 06/18, 2014, Retrieved from http://www.lamdong.gov.vn/en-us/home/dalatcity/pages/power.aspx

Le Anh Tuan and Lam Thi Thu Suu. (2013). *Quá trình phê duyệt và thực thi các dự án thủy điện ở hưu vực sông vu gia thu bồn – quảng nam và sông long đại - quảng bình*.

Le Ha. (2013). Earthquake again at song tranh 2 hydropower. 06/04, 2014, Retrieved from http://english.vietnamnet.vn/fms/society/83450/earthquake-again-at-song-tranh-2-hydropower.html

Minh Do, T., & Sharma, D. (2011). Vietnam's energy sector: A review of current energy policies and strategies. *Energy Policy, 39*(10), 5770-5777. doi:http://dx.doi.org/10.1016/j.enpol. 2011.08.010

MOIT, Ministry of Industry and Trade. (2013). *Result of hydropower dam construction survey nationwide*. Ministry of Industry and Trade: Ha Noi, Vietnam.

Nang Luong Viet Nam. (2013). Thủy điện hòa bình sản xuất trên 175 tỷ kWh. 05/16, 2014, Retrieved from http://www.nangluongvietnam.vn/news/vn/dien-luc-viet-nam/thuy-dien-hoa-binh-san-xuat-tren-175-ty-kwh.html

Nguyen Hoai Nam. (2010). Thac ba hydropower dam. 06/18, 2014, Retrieved from http://www.vncold.vn/Web/Content.aspx?distid=165

Pham Hong Giang. (2013). Opinions on hydropower development in vietnam. 06/11, 2014, Retrieved from http://www.vncold.vn/Web/Content.aspx?distid=3353

REN21. (2012). *Renewables 2012 – global status report.* Renewable Energy Policy Network for the 21st Century: France.

The management board of the project 747and 472. (2009). *The summary report of the living stability assistance for displacers in the hoa binh hydropower dam.* The management board of the project 747and 472: Hoa Binh.

The United States Energy Information Administration (USEIA). (2012). International energy statistics. 15/06, 2014, Retrieved from http://www.eia.gov/cfapps/ipdbproject/iedindex3.cfm?tid=2&pid=2&aid=2&cid=ww,&syid=1980&eyid=2011&unit=BKWH

Thien Thu. (2013). Tính toán thiệt hại sau vụ vỡ đập thủy điện ia krêl 2. 05/21, 2014, Retrieved from http://dantri.com.vn/xa-hoi/tinh-toan-thiet-hai-sau-vu-vo-dap-thuy-dien-ia-krel-2-744579.htm

Tran Tinh. (2012). Quảng trị khắc phục vụ vỡ đập thủy điện Dakrông 3. 05/20, 2014, Retrieved from http://www.vietnamplus.vn/quang-tri-khac-phuc-vu-vo-dap-thuy-dien-dakrong-3/167 011.vnp

Ty, P. H., Van Westen, A. C. M., & Zoomers, A. (2013). Compensation and resettlement policies after compulsory land acquisition for hydropower development in vietnam: Policy and practice. *Land,* 2(4), 678-704. doi:10.3390/land2040678

United Nation. (2013). Undata. 05/20, 2014, Retrieved from http://data.un.org/Explorer

World Bank. (2011). *The welfare impact of rural electrification: A reassessment of the costs and benefits', an IEG impact evaluation.* World Bank: Washington DC.

41

3   VIETNAM IN THE DEBATE ON LAND GRABBING: CONVERSION OF AGRICULTURAL LAND FOR HYDROPOWER DEVELOPMENT[10]

3.1   Introduction

Vietnam does not fit stereotypical representations of 'land grabs': large-scale acquisition of lands by foreigners are not often reported, and the country cannot be seen as easy prey for the forces of the global market. Vietnam is a strong state with sufficient means at its disposal to prevent external intrusions deemed undesirable. And yet, as this chapter aims to show, comparable processes of transfer of land to other uses and users are taking place, with important consequences for traditional users and local development. If land grabbing is a matter only of foreignisation—large-scale, cross-border land deals initiated by foreign investors (Zoomers, 2010)—then Vietnam does not warrant much attention as an object of such initiatives, although the country and its businesses are reported as actively engaged in the acquisition of land and other natural resources in neighbouring countries such as Cambodia and Laos (GRAIN, 2012; Schönweger et al., 2012). But if we are concerned with dispossession and displacement of poor people, then Vietnam offers a specific context that shows these processes of resource transfers to be much more complex and widespread that the mainstream discourse of land grabbing seems to suggest. The current land grab discussion focuses on farmland being converted from smallholders to large-scale agriculture for the production of food and biofuel for export. Much land, however, is lost in other ways, such as the expansion of urban areas into rural fringes, the creation of national parks, and infrastructure development (including hydropower projects) that take place in many parts of the less developed world (Zoomers, 2010). It is in this sense that Vietnam experiences massive transfers of land.

The ongoing industrialisation and urbanisation policy in Vietnam since the mid-1980shas accelerated the large-scale conversion of agricultural land into other uses. This process has seriously affected rural landscapes and farmers' livelihoods and has become a contested issue on the political agenda and in mass media. This chapter takes a close look at the agricultural land conversion for hydropower development in Vietnam—in order to understand the nature of the issues at stake. The data in this chapter are derived from two main sources. General information on land conversions was gathered from secondary sources such as newspapers, conference proceedings,

---

[10] *This chapter was published as* Ty, P. H., Phuc, N. Q., & and Westen, G. v. (2014). Vietnam in the debate on land grabbing: Conversion of agricultural land for urban expansion and hydropower development. In Mayke Kaag and Annelies Zoomers (Ed.), *The global land grab: Behind the hype* (pp. 135-151). London and New York: ZED.

academic and statistical publications, legal documents on land policy, and unpublished documents. To determine the differences between Land Law 2003 and Land Law 2013, we made a comparison between two documents by using the menu Review in the Microsoft word 2010. From that, we focused on the analysis of new points of the new Land Law 2013.

The remainder of the chapter is organised into four sections. The first section presents an overview of agricultural land acquisitions and conversions in recent years. The second section focuses on explaining why agricultural land has been converted to different purposes in Vietnam The third section concentrates on the case of land conversion for hydropower development, in which it looks at the historical of policy on land conversion at different periods, describes the current land conversion mechanism for hydropower dam construction, and identify shortcomings and consequences caused by the conversion of land to hydropower dams. The last section concludes and discusses the material presented.

3.2   Agricultural land conversion in Vietnam: an overview

GRAIN, a Barcelona-based NGO engaged in the struggle for securing land rights for poor people, maintains a website listing media reports on cross-border land deals. Although the listings do not claim to be authoritative (i.e. tested and confirmed), the website has the merit of drawing attention to what is increasingly perceived as a major issue and offers a quick overview of what land transfers may be taking place around the world. Table 3-1 presents a summary with respect to the position of Vietnam in international land deals. While some foreign acquisitions in Vietnam are reported – largely related to forestry – the main thrust is clearly a different one: Vietnamese investors also emerge as important 'perpetrators' in their own right, picking up extensive holdings for cash crops in less densely populated neighbouring countries.

As we mentioned in the previous chapter, Vietnam has nearly completed their hydropower development plan in recent years. Many hydropower companies have become the owners and contractors of scale-large project not only in Vietnam but also in other countries. For example, Vietnam Electricity Group (EVN) established the International EVN in Cambodia and Laos in 2007 in order to expand their investment to those countries. Since 2010, many hydropower projects have been promoted in neighbouring countries, including Laos and Cambodia and there is an increase in coming years as mentioned in the electricity development plan of the Vietnamese Governement for the period between 2011 and 2015, and toward 2030. Currently, several hydropower dams are being constructed in Laos and Cambodia, including Xekaman 1, Xekaman 3, and Nam Mo in Laos; and Se San 2 in Cambodia. From 2015 to 2020, six hydropower dams will be constructed in Laos and Cambodia. Most electricity produced by hydropower plants in Laos and Cambodia will be exported to Vietnam (MOIT, 2011). Thus, the hunger for electricity is a driver for the expansion of land acquisition for hydropower development of Vietnam in other countries. It would be misleading to conclude, however, that land acquisitions and related displacements are not an issue in Vietnam.

Table 3-1 Vietnam land deals in other countries and foreign deals in Vietnam. *Source*: International Land Coalition: Land Matrix, http://www.landmatrix.org/get-the-idea/web-transnational-deals/(accessed 6 July 2013)

Type of land deal	Types of investment	Area (ha)
Land deals in other countries		
Cambodia	Agriculture, Industry	9,380
Cambodia	Agriculture	6,436
Cambodia	Agriculture	2,361
Cambodia	Agriculture	9,784
Cambodia	Agriculture	8,000
Cambodia	Agriculture	6,891
Cambodia	Agriculture	8,100
Cambodia	Agriculture	7,600
Cambodia	Agriculture	7,560
Cambodia	Agriculture	2,502
Lao	Agriculture, Forestry, Other	10,000
Lao	Agriculture	10,000
Lao	Agriculture	10,000
Sierra Leone	Agriculture	110
Cambodia	Agriculture	6,155
Cambodia	Agriculture	7,000
Cambodia	Agriculture	9,014
Cambodia	Agriculture	9,656
Cambodia	Agriculture	9,614
Cambodia	Agriculture	9,773
Cambodia	Agriculture	7,900
Cambodia	Agriculture	7,591
Cambodia	Agriculture	1,900
Cambodia	Agriculture	4,889
Cambodia	Agriculture	9,785
Cambodia	Agriculture	6,695

Type of land deal	Types of investment	Area (ha)
Cambodia	Agriculture	5,080
Cambodia	Agriculture	5,095
Cambodia	Agriculture	2,183
Cambodia	Agriculture	7,289
Cambodia	Agriculture	7,972
Total area (ha)		**254,392**
Foreign land deals in Vietnam		
Cambodia	Agriculture	5,345
China	Agriculture	10,000
Hong Kong	Forestry	63,000
Japan	Conservation, Forestry	309
Israel	Agriculture	2,500
Switzerland	Agriculture	unknown
Hong Kong	Forestry	100,000
Hong Kong	Forestry	21,000
Hong Kong	Forestry	70,000
Hong Kong	Forestry	30,000
Hong Kong	Forestry	65,000
Japan, China, Hong Kong	Forestry	3,500
Total area (ha)		**370,654**

On the contrary, the emphasis in the land grab debate on foreignization of land and on agricultural land obscures the fact that other conversions are taking place on a massive scale, with often similar consequences for rural populations. For this we need to take a closer look at what happens. Vietnam conducts a nationwide land use census every five years, and similar data are also identified every year in each province, district and commune. Land use changes over the last decade are listed in Table 3-2. It can be seen that the total land area of the country increased somewhat, as a result of new measurements. Over the decade, there was a considerable drop in the area of unused land. Much of this consisted of upland areas subject to reforestation programmes and, as a result, reclassified as forest land, a subcategory of agricultural land. Non-agricultural land uses increased rapidly, from 1.7 million hectares in 1990 to 1.9 million hectares in 2000 and 3.5 million hectares in 2009 (GSO, 2010). In relative terms, expansion of residential and special-use land was the most spectacular change observed. Special-use land increased by approximately 150,000 hectares per year, and the increase in residential areas amounted to approximately 21,000 hectares each year.

However, these aggregate statistics to some extent hide what is actually happening in terms of land conversions.

Table 3-2 Land use change between 2000 and 2009 (1,000 ha). *Source: GSO, 2010*

Land use types	Year 2000		Year 2009		Change	
	Area (ha)	per cent	Area (ha)	per cent	Area (ha)	per cent
Total area	32,924.1	100.0	33,105.1	100.0	181.0	
Agricultural land	20,920.8	63.5	25,127.3	75.9	4,206.5	12.4
Special-use land[11]	1,532.8	4.7	2,835.3	8.6	1,302.5	3.9
Residential land	443.2	1.3	633.9	1.9	190.7	0.6
Unused land	10,027.3	30.5	4,508.6	13.6	-5,518.7	-168

More detailed information shows that most additions to special and residential land use types (non-agricultural land) have been converted from agricultural land, especially rice fields and dry croplands. However, there is no systematic reporting on the size and nature of land conversions (Nguyen Van Suu, 2009). Therefore, we have to rely on a range of different bits and pieces of information scattered over several sources. It is estimated that over 80 percent of new urban and industrial developments affect agricultural land (Bui Ngoc Thanh, 2009). Mai Thanh (2009) claimed that in the period 1995–2005, over 766,000 hectares of agricultural land had been converted to urban and industrial use by the central government. This accounted for 4 per cent of the total agricultural land area of Vietnam. While this suggests that the scale of conversion is relatively limited, data from the General Statistics Office present another view: the area of rice production land, for instance, decreased from 6.7 million hectares in 1995 to 4.08 million hectares in 2009, while the non-agricultural land uses[12] increased rapidly, from 1.7 million hectares in 1990 to 3.5 million hectares in 2009 (GSO 1996, 2010). In 2009, Báo Quân đội Nhân dân (Vietnam People's Army Newspaper) reported that approximately 59,000hectares of rice land were being

[11] According to the land use classification system of the Land Law 2003, special-use land includes land used for construction, transportation, irrigation, public structures, commercial and non-agricultural production purposes, special-use water bodies (hydropower, irrigation reservoirs, waterways), cemeteries and religious use, and other non-agricultural land
[12] According to the land use classification system of the Land Law 2003, unused land comprises land for which the purpose of land use has not yet been determined and not allocated to and land users

appropriated each year for non-agricultural purposes. The central government founded 228 industrial, economic and high-tech zones between 1991 and 2010, using 49,330 hectares of land, excluding a variety of small-to-medium-sized industrial areas managed by provincial authorities. The total area of industrial zones is expected to be 70,000 hectares by 2015 and 80,000 hectares by 2020 (Nguyen Van Suu, 2009, p. 109). On the national scale, land for infrastructure development increased from 437,963 hectares in2000 to 1.08 million hectares in 2009 (GSO, 2001 and, 2010).

Urban expansion entails more than just the construction of residential areas, business parks and urban infrastructure. Large areas have also been used for amenities, such as the 104 golf courses created between 2003 and 2008 alone, occupying on average 471 hectares each and thus consuming 49,000 hectares in all (Duy Huu, 2008). In addition, many coastal lands, including protected forests and farmlands, have been allocated to domestic and foreign companies to build resorts. Da Nang city is an example, where dozens of luxury resorts have opened along the coast, including the Empire Residences and Resort (51 hectares) and Da Nang Beach Resort (260 hectares). A particular source of land loss in Vietnam is the construction of hydropower dams. This results from a policy aiming to meet rapidly increasing demand for electricity while limiting the import bill for fuels. For instance, Hoa Binh hydropower dam was constructed in central Vietnam in 1979 and flooded 75,000 hectares of land, including 11,000 hectares of agricultural land; 11,141 households, with 89,720 people, were removed from the flooded areas (Institute of Development, 2010).

Farmland loss for urban and industrial expansion is concentrated particularly in some sixteen provinces, which include the largest cities. Most land conversion takes place near very large cities such as Hanoi and Ho Chi Minh City, where prominent new urban developments have mushroomed. Increasingly similar processes also occur in secondary cities such as Vinh, Hue, Binh Duong, Dong Nai, Vinh Phuc, Hai Duong, Da Nang and others. It will be clear that such massive reallocations of land to new uses and users must affect many people. On the national scale, land conversion influenced the livelihoods of 627,000 households and 2.5 million people between 2003and 2008 (Mai Thanh, 2009). In the capital region of Hanoi, urban expansion between 2000 and 2010 entailed the conversion of 11,000 hectares for 1,736 projects. This resulted in the loss of traditional employment for some 150,000 farmers (Nguyen Van Suu, 2009). Hong Minh (2005) reported that in just five years, from 2000 to 2004, 5,496 hectares were used for 957 projects, critically impacting the life and employment of 138,291 households, of which 41,000 were classified as agricultural households.

Drivers of land conversion

Changes in land use and in claims on land enjoyed by different social groups are thus observed on a massive scale in Vietnam. While part of this concerns rural transformations such as the establishment of large-scale commercial agriculture and reforestation programmes, the main thrust is of a different nature and marks the structural transformation of Vietnam from a predominantly rural economy and society to an urban one based on manufacturing and service industries. This transformation is linked to various domestic and external processes. External forces can be conveniently labelled as globalization. This entails Vietnam's progressive integration into the global economy as a suitable location for a range of globally integrated production activities and ways of organizing the economy, guided essentially by the prescriptions of neo-liberalism. This globalization is reflected in an increasingly open economy, with prominent roles for foreign investment and export production, but also in the adoption of corresponding consumption and lifestyle elements – as symbolized by the golf courses and resorts mentioned above. Domestically, the globalization drive has been activated by the adoption of a comprehensive set of policies known as Doi Moi[3] (renovation policy) since 1986. In spite of continued adherence to the socialist foundations of the Vietnamese state, this reform process has effectively aligned Vietnam's institutional framework with those of the main capitalist economies. In spatial terms, globalization and economic transformation (industrialization) are reflected in large-scale urbanization, favouring especially those localities that have advantages in terms of connecting with the world at large. A major component of the transformation, as observed, concerns land: how it is used, who uses it, and what is the institutional framework of laws and procedures that guide its access and use.

Economic and social transformations: The transformation of economic and social conditions has put particular pressures on land, while its availability is limited because most land is allocated to individuals, family households and organizations for certain periods of time (Ngo Viet Hung, 2007; Nguyen Van Suu, 2009). Conversion of agricultural land to urban uses and users is seen as a major requirement to facilitate the realization of Vietnam's transformation into an urban and industrial society, and hence the institutions pertaining to land had to be modified accordingly. Four factors that define these changes will be briefly reviewed: political and economic transformation, urbanization, the emergence of 'post-productivist's green spaces, and changes in land policy.

Political and economic structure transformation: Economics and politics are the main drivers of the transformation process, with the change from a centrally planned towards a market-oriented model of economic governance. Accordingly, central state dominance in economic and social affairs has given way to a new division of administrative responsibilities between central and local authorities. The top-down approach has been replaced by more flexible strategic planning in a number of fields. New legislation, such as the Enterprise Law (2000), the Land Law (2003) and the Investment Law (2005), has encouraged economic growth, export production and foreign investment. Vietnam's GDP growth averaged 7 per cent per year between 2006 and 2010 (GSO, 2010). Foreign investment reached US$29.4 billion and exports rose from US$23 billion (2001–05) to US$56 billion in the period 2006–10 (ibid.). The industrial sector plays an increasingly important role in the economy, contributing 22.6 per cent of GDP in 1990 and 41 per cent by 2010 (GSO, 2011). The transformation of political and economic structures has also significantly contributed to changes in the relationship between the central and local authorities in a number of fields, such as socio-economic development strategy, natural resources management and urban development planning.

Urbanization: In Vietnam, urbanization has rapidly increased since the national economy began to be integrated with the global economy. There was a marked increase in the 2000s, with 19.4 per cent of the population living in urban centres in 2001 (Coulthart et al., 2006), compared with 33 per cent in 2010 (UNDESA, 2010). At present, at least one million people are added annually to Vietnam's urban areas; the current annual 3 per cent increase in urbanization means that the urban population will constitute around 41 per cent of the total in2030. Vietnam's future economic growth will depend on its ability to develop competitive, market-driven industrial and service sectors. These are primarily urban-based activities (Coulthart et al., 2006). The cities and towns account for approximately 70 per cent of total economic output (ibid.). The large cities of Ho Chi Minh City, Hanoi, Hai Phong and Da Nang are the focal points in the transition to a market-oriented economy (Phan Xuan Nam et al., 2000).

In the last two decades, the central government has introduced reforms that have effected urban development. The first reform was set down in the Orientation Master Plan for Urban Development to 2020, adopted in 1998 (Bo Xay dựng, 1999). This explicitly addresses urbanization by designating a hierarchy of urban settlements (Coulthart et al., 2006). However, the most important policy change was the introduction of the new Construction Law in 2004. Its main feature is increased decentralization to the three lower tiers of government– provinces, districts and communes or wards – in preparing spatial plans (still subject, however, to approval by central government).In practice, most FDI is directed toward cities owing to their competitive advantage in terms of labour market, transportation costs and infrastructure. As mentioned above, a large number of industrial sites and export processing zones have been established in and near cities since 1991.

Emergence of green-space consumption trends: Recent surveys have shown that the quality of life has declined in several cities in Vietnam, as city planning does not necessarily meet citizens' needs (Ngo Tho Hung, 2010). One possible reason is the decline of green spaces within urban areas, although other factors such as healthcare, education systems and entertainment are also important. In addition, air pollution is becoming a serious issue in the very large cities such as Hanoi and Ho Chi Minh City. Most urban air pollution originates from traffic, manufacturing and domestic cooking. It is estimated that 70 per cent of urban air pollution comes from vehicles, in particular motorbikes (ibid.). A healthy living space includes green urban attributes as well as socio-economic and cultural services (Mahmoud & El-Sayed, 2011).In recent years many Vietnamese, especially the nouveaux riches, have developed a preference for leafy residential areas where they can relax and recover from daily stress (Waibel, 2006). The emergence of such new urban areas is a response to aspirations of people who have medium and higher incomes and aspire to lifestyles that differ from others. By moving into these new residential areas, such people expect to adopt an upscale Western lifestyle in secure, orderly neighbourhoods, with fresh air, green spaces and comfort. In these residential spaces, they feel part of a globalizing modern society in a setting that meets international standards (ibid. 2006).

The model of new eco-urban areas has been promoted in Vietnam (Labbe & Boudreau 2011) since the late 1990s. These urban spaces are built on agricultural land in peri-urban areas. Within these areas, high-rise apartment blocks, villas and commercial and office spaces have been built, along with education, health and sports facilities and parks. By the end of 2010, 633 new urban areas had been realized all over Vietnam, with a total area of 103,243 hectares (An Bien Real estate, 2011). Prominent new urban developments are the Sai Gon South New Urban Area (Phu My Hung) in Ho Chi Minh City, New City in Binh Duong Province, and Ciputra International City in Hanoi. Saigon South is being developed by Taiwanese investors and occupies 3,300 hectares of former wetlands. Its population is projected to be between 500,000 and 1 million residents by 2020. The New City in Binh Duong covers 1,000 hectares and was designed by the National University of Singapore. Ciputra International City near the West Lake of Hanoi is an Indonesian investment of US$2.1 billion. It occupies 405 hectares of land with villas, commercial centres and high-rise blocks of seventeen to twenty floors. A 120-square-metre apartment can be sold for about US$100,000 (Waibel, 2006, p. 46). In this urban residential space, inhabitants control and to some extent even shape their own territory, which may be interpreted as a privatization of part of urban space (ibid.: 47).

3.3 Change of land conversion, compensation, and resettlement policy in Vietnam

Since 1953, a large number of legal documents have been issued to manage land recovery, compensation, displacement and resettlement, and livelihood development for displaced people in Vietnam. To understand these policy changes, it is necessary to characterize the evolution of property rights of displaced people, including land and property ownership types, land use rights, land management authorities, participation mechanism in decision-making processes, compensation and resettlement solutions, and assistance regulations for stabilizing displaced people's lives and for developing their livelihood after displacement. There are three key stages, which convey the most important changes in the legal framework and its implementation in practice.

Before the Constitution 1980: The first land-recovery policy was regulated by the Agrarian Reform Law 1953 to take possession of land from French colonial and feudal landlords (địa chủ) without compensation. Compensation of cash or government cheques for property losses were only provided in cases where land was acquiring from other farmers. Land appropriation mainly served political purposes and not social-economic development. During this period, governmental non-compensated land appropriation mainly targeted landlords, and, therefore, they lost the most land from this reform. Recognizing these mistakes, the government issued a corrective regulation, Decision 51/CT-TW, in 1956 to return land to landlords and to regulate compensation for other landowners' losses. After the implementation of agrarian reforms, the Constitution 1959 was formulated, in which it recognized three forms of land ownership, including state, individual, and cooperative ownerships. However, land resources typically belonged to the cooperatives in practice, with each farmer a member. In 1959, Circular 151/TTg was issued to control land acquisition for the purpose of government-managed construction. According to this circular, the principle of land recovery was to ensure sufficient land for state construction and to pay attention to compensation for landowners' losses. Negotiations among land-losers, cooperatives, and provincial committees were intended to provide estimation for compensation.

Land loss was compensated with other land, and property damage was paid by crop productivity unit per land area; there was no need to elaborate a plan for compensation and resettlement. In 1970, Decree 1792/TTg was adopted to regulate compensation for development activities, including opening new economic zones and urban expansion projects. Nonetheless, only property loss was compensated corresponding to the price of rice, while loss of land was not recompensed.

Between 1980 and the Constitution 1992: In 1980, a new constitution was born which made tremendous changes in land ownership compared with the 1959. There was no longer private and cooperative landownership. It was replaced by a single type of land ownership: land was owned by the entire people under the management of the state. The state held the right to allocate land to land users and could take land back for the sake of national development and security. Land losers could receive compensation for property losses but land was not recompensed because it belonged to the state. Land Law 1987, the first of its kind, was adopted based on the Constitution 1980. In term of this law, the land was still possessed by the state, but the government allocated land to more actors, such as state agriculture and forest companies, cooperatives, military, administration offices, social organizations, and individuals to use long-term. This was the first time that land management activities were clarified and categorized, in which land acquisition was one of the land management components. The purpose of land acquisition was expanded not only for state purposes but also for social objectives. More importantly, land users had three land use rights, including transferring, exchanging, and compensation for land loss due to government re-acquisition. The government also granted land use right certificates (Red Book) and house ownership certificates to land users, so land rights were more secure than before. Nevertheless, compensation methods were still vague with supports for land losers concentrated solely on agricultural and forestland when these lands were converted to other purposes. Several compensation methods were formulated to identify the price of agriculture and forest land.

From the Constitution 1992 to Land Law 2003: The 1992 Constitution laid the foundation for socio-economic reform policy. The economy of Vietnam was transferred from a subsidized system to market-orientated economy. However, state land ownership remained as in the previous Constitution. Based on the 1992 Constitution, the 1993 land law was approved. This law expanded the previous three land use rights to seven rights, including rights to exchange, transfer, lease, inheritance, mortgage, land investment as capital, and to be compensated due to state land acquisition. According to the law, compensation must be estimated at the market price when the government appropriated land for purposes of national security, national, and public benefits. This was the first time that land market was officially stated in law. Land prices were determined by government for different regions and periods. Another improvement was the extension of land use for a longer duration. The central government also decentralized the rights to lower authorities to manage land allocation and acquisition. Following this, many regulations on compensation and resettlement were elaborated, as well.

Decree 87/CP, issued in 1994, stipulated land price valuation, and land price was estimated from a maximum-to-minimum limit in terms of different criteria, including location, infrastructure, and urban types. Decree 90/CP, issued in September 1994, expanded the conditions for land compensation. Under this regulation, lands without Red Book were also considered for compensation if they had no dispute with other land users and documents to show the progenitor of land ownership. Nevertheless, the decree did not take into consideration any kind of resettlement program and livelihood support for land losers. Therefore, in 1998, Decree 22/1998/NĐ-CP was issued to address this issue.

Under this decree, all projects were required to prepare a resettlement plan for displaced people. This was the first time that resettlement planning was officially regulated by government and local authorities. Compensation was categorized into five types, including compensation for land and property losses, support for displaced people to stabilize after resettlement, costs for job training, and costs for local organizations to implement compensation, resettlement, and land clearance. Methods of estimating land price for compensation were further elaborated; the profit of land was taken into account. Payment could be in cash or land, depending on the preferences of displaced people. However, support policies only paid attention to early years of resettlement but not for livelihood restoration in the long-term.

After the Land law 2003: The adoption of Land Law 2003 did not make any change in land ownership but increased from seven to ten rights land use rights for land users, including the rights of exchange, transfer, inheritance, mortgage, lease, sublease, guarantee, as well as land pooling, donating or gifting, and compensation. Furthermore, more land users were recognized in the Land Law, including community, religious and foreign organizations, and ex-pat Vietnamese. The land market was officially legalized to operate the market in a transparent way because the land market was privately transacted between land users without using real-estate companies to complete deals. More importantly, it was stated for the first time in this Land Law that land recovery was for the purpose of economic development, such as hydropower development, urbanization projects, industrial zones, and resorts and tourism, as well as ODA projects. This was also the first time that land without Red Book was eligible to be compensated in cash or other lands. The government also declared that "the resettlement plan must support displaced people in new resettlement sites to have at least equal or better living conditions than old places." This was quite similar to the slogan of World Bank (Ty et al., 2013). In cases of agricultural land-loss of over 70%, farmers could receive the payment in cash and access to support programs, food security, and job replacement support.

Following this law, further legal documents were adopted to direct the implementation. Decree 188/2004/NĐ-CP, formulated in 2004, regulated the estimation of land price compensation, in which the market price was clearly defined

(See Ty et al., 2013, p. 683). After this, the Ministry of Finance promulgated Circular 114/2004/TT-BTC to instruct land price valuation methods of using market information and income data to determine land price for compensation and land taxation. For land recovery, Decree 197/2004/NĐ-CP was adopted to replace Decree 22/1998/NĐ-CP, in which the implementation of land recovery, compensation, and resettlement was authorized to the Provincial People's Committee (PPC). The decree opened more doors for local authorities to decide on resettlement options, depending on their local contexts. In 2006, Decree 11/2006/ND-CP was issued to require PPC to annually update the land price for land compensation and land tax. Additionally, poor households were provided more supports to stabilize their lives for a period of 3 to 10 years, if their land was appropriated. In 2007, Decree 123/2007/NĐ-CP authorized PPC to formulate a land-price framework of all land use types for the sake of land compensation, land registration, land tax, and land transaction fees. Decree 84/2007/ND-CP regulated the procedure of compensation and resettlement when the state recovered land, including a ten-step process. In 2009, Decree 69/2009/ND-CP modified previous decrees, of which the newest instruction was the establishment of the Land Development Fund at the provincial level on the basis of revenue from land. The compensation and support program was also separated to implement.

For land acquisition for hydropower construction, Decree 34/2010/QD-TTg, issued in 2010, laid the first foundation for the improvement of compensation, displacement, assistance, and resettlement for irrigation and hydropower dam projects. The first component of displacement and resettlement was compensation by lump-sum cash payment for losses of land and other properties. The second component was assistance for resettlement, including supports for food, health care, education, fuel, electricity, and agriculture production, which were extended from one to two years. The principle of assistance aimed mainly to recover agricultural production and forest livelihoods. Resettlers could also receive technical training and financial assistance to build crop and livestock models. In this decision, responsibility for implementing compensation, resettlement, and livelihood program was clearly defined for authorities from central to local level.

Changes in the new Land Law 2013: A new land law was adopted by the National Assembly November 29, 2013. This law was prepared by the drafting committee, including the General Director of the General Department of Land Administration, representatives of Ministry of Agriculture and Rural Development, Ministry of Justice, Ministry of Finance, Ministry of Construction, Ministry of Transport, the State Bank of Vietnam, Ministry of Planning and Investment, and the Research Unit of the Prime Minister. A draft was publicized to collect people's opinion cross the country. However, the adopted law did not make any significant reform to land ownership and management; it tended only to reinforce more clearly the rights and responsibilities of state management agencies and land users. Government and local authority retained sovereignty over land acquisition for national security and defence and socio-economic development projects for the national and public interests. However, types of land acquisition are obviously specified (see Article 62, 63), in which hydroelectricity is assigned to the electricity development projects, which required applying a compulsory land conversion mechanism. A strict land use and management monitoring system was established from central to local level, including supervision by the National Assembly, provincial and district people's councils, and members of Vietnam Father Front (See Article 198). More importantly, the supervision of all citizens, implying civil society organizations and individuals, was newly regulated; they had the right to carry out the supervision over the land management and use of the state, local authority, and other land users themselves or their associations, which implied the involvement of civil society organizations, though not clearly stated (See Article 199). For land acquisition, this law aimed to increase the accountability of the state and local agencies (See Article 13-19). Another improvement was that the new land law aimed to improve the land use planning transparency(Article 43); land conversion process (Article 69 and 199); compensation, support, and resettlement aims (Article 86); and to expand the participation of land losers.

Furthermore, although the state allowed the consultancy companies to participate in land valuation to identify the specific prices for land compensation, price was still regulated by state and local authorities. Customary rights over land are not mentioned in this law, so the loss of customary land (reclaimed land, slash and burn-based agricultural land, and communal land) might be not compensated. Compared to Decree 34/2010/QD-TTg, applied for the case of land acquisition for hydropower projects, there were few innovative points. Few elements in Decree 34/2010/QD-TTg were upgraded in this new law. More importantly, the "No" option for land losers to refuse the project was not taken into account in the process of land conversion. It was not required to obtain the acceptance of affected person before making decisions on land acquisition. Thus, principles, like Free, Prior, Informed Consent (FPIC), are partly integrated into the new law. However, these regulations still lacked of specific mechanisms to effectively implement in practice, which may influence its efficiency and feasibility. This did not seem to be an improvement, as it was not as comprehensive as the Land Acquisition Act of India of the same year, where 70% of land losers in the case of Public Private Partnership projects and 80% in the case of private companies shall consent prior to the process of acquisition (Pawar, 2014). Thus, we can predict that the problems of land acquisition in Vietnam will be remained as previously and even more severely.

An example of mechanism for land conversion, compensation, and resettlement: The following example was extracted from Decision 18/2011/QĐ-UBND of Thua Thien Hue province. The mechanism of compulsory land recovery, compensation, support, and resettlement is presented in Box 3-1. It can be understood that the mechanism of land acquisition, compensation, and resettlement was imposed by a top-down approach. The decisions are made by most state agencies, and therefore affected people do not have any opportunity and space to intervene or negotiate with state agencies for compensation prices. They can only claim their rights once the land appropriation has already occurred; prior to appropriation, they have no right to influence the decisions of state and local governments. This process partly reflects the land grab mechanism by the state, local governments, and private investors. The way of implementing this type of compulsory land acquisition has led to serious consequences for the acquired landholders. The following consequences (see Box 3.1) caused by land acquisition for hydroelectric dam construction will illustrate weaknesses of this land acquisition system.

Box 3-1 Land conversion mechanism of Thua Thien Hue province in 2011. *Source: http://vbpl.thuathienhue.gov.vn/default.asp*

Phase I: Before making decisions on land recovery and allocation to investors
Step 1: Investors submit the investment project application to the competent authorities, depending on the type of projects.
Step 2: The competent authorities consult with functional departments to introduce suitable locations for investors. If an agreement is made, the Provincial People's Committee (PPC) accepts the investment undertaking
Step 3: Next, investors must prepare the Master Plan for Compensation, Support and Resettlement and the plan for job training and replacement for farmers.
Step 4: Following that, PPC notifies or authorizes District People's Committee (DPC) to notify of the land acquisition to the Commune People's Committee and affected people. The notification must be propagated in the local media, such radio, TV, and posted at public areas.
Phase II: Approval and implementation of decisions on land recovery, compensation, support, and resettlement
Step 5: Approval of land recovery: PPC make decisions on land recovery, allocation or lease to investors
Step 6: Land and property loss inventory is conducted together with households, leaders of the Commune People's Committee (CPC), cadastral staffs, representatives of Board on compensation, support, and resettlement (BCSR), and investors. After

the survey, all representatives and land losers must sign in the minutes if in agreement with the survey; if not in agreement, opinions provided by land losers must record in the minutes.

Step 7: BCSR prepares a detailed plan for compensation, support, and resettlement, including compensation proposal, displacement and resettlement plan, and job replacement and training program for farmers if the acquisition is subject to agricultural production land

Step 8: BCSR is responsible to announce publicly the detailed proposal for a period of 20 days at the office of CPC and at public places close to affected communities whose land is recovered, for the purpose of collecting feedback from them subject to land acquisition decisions, compensation, and resettlement.

Step 9: After receiving feedback from affected people, BCSR must to prepare a report of people's opinions and submit a modified plan to the competent agency at the same level for approval.

Step 10: Following that, PPC makes a decision to approve or disapprove the compensation, support, and resettlement plan and informs the affected people of the approved plan.

Step 11: Implement the payment for compensation and support and arrangement of the resettlement plan must be begun within 45 days after PPC approves the proposal.

Step 12: After receiving compensation, land losers must abandon their land and move out. BCSP prepares a minute to be signed by those displaced and CPC. If land losers do not leave, they will receive an eviction notice. If, after 15 days of this notice, they continue to reside on the land, they will be forced to leave by police and BCSR.

Phase III: Affected people implement the rights of denunciations, complaints, and lawsuits

Denunciation: If affected persons discover any agency, organization, or individuals who has violated, threatened, or damaged the rights and legitimate interests of the state, citizens, agencies, and organizations, these affected persons reserve the right to inform the responsible authorities in term of the Accusation Law.

Complaints and lawsuits: If land losers disagree with the decision of land acquisition, compensation, support, or resettlement, they can send complaints or lawsuits against administrative decisions and implementation of responsible agencies or staffs in accordance with the Denunciation Law and Civil Procedure Code.

It can be understood that the mechanism of land acquisition, compensation, and resettlement was imposed by the top-down approach. Decisions made by most state agencies affected most those people who do not have any opportunity and space to intervene or negotiate. Accordingly, they may only claim their rights when once appropriation has occurred. Prior to appropriation, they have no right to change the decisions of state and local governments. This process partly reflects the land grab mechanism by the state, local governments, and private investors. The way of implementing this type of compulsory land acquisition has led to serious consequences for the acquired landholders. The following consequences caused by land acquisition for hydroelectric dam construction will illustrate weaknesses of this land acquisition system.

3.4   Shortcomings and consequences of land conversion to hydropower development

Intensive displacement: Before 1990, more than 120,000 people were displaced for the Thac Ba hydropower dam, and approximately 90,000 for the Hoa Binh hydropower project. In the mid-1990s, 60,000 people had to make way for the Ham Thuan–Da Mi dam and over 24,000 for the Yali dam respectively. By the end of the 1990s, over 400,000 people had been displaced owing to hydropower projects in Vietnam, according to the Institute of Strategy and Policy on Resources and Environment (2009). Another source (Department of Co-operatives and Rural Development, 2007) reports that between 1995 and 2009, 49,000 households were affected by the construction of over 20 large-scale hydropower dams (stations with capacity over 100 MW), claiming 80,000 ha of land. For example, Son La hydropower dam displaced over 90,000 people from 160 settlements (Consultation Company on Electricity I, 2007). Less spectacularly, many small hydropower dams also displaced a large number of people—for example, Nam Na dam (2,325 persons) and Khe Bo dam (3,482 persons). A salient feature of these displacements is that some 90 per cent of the affected people belong to minority ethnic groups living in mountainous areas (see also Nuijen, Prachvuthy and Van Westen on Cambodia, this volume). Several dams actually displaced only members of ethnic minority peoples such as Hoi Quang, Nho Que 3, and Bac Me (Consultation Company on Electricity I, 2007; Stockholm Environment Institute, 2007; NIAAP, 2008; CODE, 2010). To 2013, the construction of hydropower dams has displaced about 60,000 households with 240,000 people. From 2013 to 2030, over 60,000 people will be displaced for 21 large hydropower projects. Thus, approximately 300,000 people (about 0.3% of total population) are displaced for the entire plan of hydropower development of Vietnam. In which, 90% of them are poor ethnic minority groups relying on forest and agriculture livelihoods and are facing many challenges to sustain their lives after resettlement (MOIT, 2011 and 2013; Ty et al., 2013; Dao, 2010; Hang, 2013). In addition, over 620,000 hectares of land have been lost so far, in which small hydropower dams accounts for 6% of total loss, in which forest and agricultural land accounts the most (MOIT, 2013).

Dissatisfaction of resettlement: Displacement of people requires providing other places to resettle, therefore the displacement and resettlement is often designed in the same program called "the detailed plan of displacement and resettlement". The program is designed by the district authority in collaboration with the investor, in which the project developer has to pay for the program. There are three types of resettlement, including the on-spot (di ven), the integrated (xen ghep), and the centralized (tap trung). The first type moves displaced people surrounding the reservoir of dams, the second relocates people into different communities existing in other places, and the last one gathers all displaced people in a new centralized area. Most projects apply the third type. It is a new place where the district authority recovers land from other landholders to construct a new resettlement for the displaced.

However, most resettlers don't satisfy with the new living conditions in the relocation site. The first reason is that agricultural production land is much smaller than the previous land in the origin, while displacers are typically farmers. Even after three years of resettlement, 99 households did not receive the agricultural production as committed by the investor (Ho Moong commune people's committee, 2009). In the study of (CODE, 2010) in four large dams, including Hoa Binh, Ban Ve, Yaly, and Tuyen Quang, about 86 percent of respondents confirmed that their land areas in the new destination were far narrower than the origin. Only 7 percent of them received more land than before. Moreover, 77 per cent of respondents disappointed with the very low quality of land in the relocation region, so a lot of people did not accept to live in the resettlement site of the Pleikrong dam. The resettlement site is also not suitable to their traditional culture and religion. Most ethnic minority groups practiced the slash and burn cultivation, they moved around mountainous areas in the past and selected the most appropriate region to settle down where the area used to rich soil and near the forest since their religion was much associated with the natural forest. But at the new destination, all are gathered in the densely population, far from the forest, water courses, and grass land. They can no longer access to those common properties to practice their traditional festivals, and can't use resources from these common pool resources, such as hunting, fishing, collection of bee honey, rattan, and fruits in the forest. Additionally, they could not find non-farm works in the new environment because they depend mostly on agricultural production activities. Therefore, over 60 per cent strongly displeased with the new destination in 4 cases.

Inadequate compensation: All hydropower dams implement compulsory land recovery programs, people whose lands are situated in the reservoir of dams have to move out and return lands to the State. And then, the State leases or allocates lands to the investor of hydropower dam. It is very different from the cases that people voluntarily contribute lands to several other projects, such as national road construction, public structures, and social development projects. In forced displacement, affected people consider the compensation for the land and property loss seriously. In addition to the improvement of compensation policy so far, such as higher compensation for agricultural land, property losses compensated as the market prices, and higher involvement of affected households in the property loss surveys, the compensation for land often is far lower than the market price. For instance, 95 percent of land losers in the Tuyen Quang dam under-evaluated the compensation price unit and 94 percent in the Yaly dam. Over 91 percent of respondents ascertained that compensation for annual trees and crops were very low in the Tuyen Quang dam, similarly 94% in the Ban Ve and 91% in the Yaly dams. Therefore, they did not afford to buy other lands. For example, people received about 37 million VND per hectare for agriculture land loss in Yaly hydropower dam, but the market price at that time was 43 million VND; the compensation for agriculture land was 26 million VND in the Pleikrong dam but it was 40 million VND. Furthermore, land and property losses were compensated insufficiently, many were not estimated. Over 50 percents of people in the Yaly dam, 34% in the Ban Ve, 60% in the Hoa Binh, and 67% in the Tuyen Quang dam confirmed the inadequacy of the land and property loss surveys. As a consequence, 88 percent of respondents in 4 projects strongly discontented with the compensation policies (CODE, 2010).

Limitations of livelihood restoration and job replacement programs: Besides the compensation and resettlement program, the assistance program for living and agriculture production is also compulsory once upon land recovery for hydropower dam construction. Nonetheless, it only provides assistances in the short-term, such as rice, fuel, health care and school fees, and supports for festivals before living in the new area, agricultural production tools, and job training. It is also very different among projects. Most projects support rice for a year, some provide two years like the Son La and Ban Ve dams. However, most households receive assistances by cash, so people have to buy higher prices of rice and other things. Therefore, about 87 percent of the displaced under-valued these assistances in three hydropower dams because they were little and instable for their new living conditions. Additionally, nearly 50 percent did not appreciate the agricultural extension services because it did not help in improving their agricultural production (CODE, 2010). The job replacement program also does not work for the ethnic farmers whose education is low. The resettlement site is located also in the agricultural lands and rural areas, but lands do not belong to the resettlers. Hence, they cannot go on practicing agriculture activities on a very small piece of land. Other jobs are not available for not only resettlers but also the local people. In spite of following the program, most people want to receive job assistance in cash.

Furthermore, the most important livelihood source of farmers is the arable land, but resettlers haven't received sufficiently as the project's commitment before. For example, the household survey in the A luoi hydropower company shows that the investor haven't given 0.5 arable land displaced people as committed. Generally, the most difficult problems for resettlers' livelihood restoration are lack of arable land and job replacement. Even the dam constructed in 30 years ago, resettlers are still facing the same problems, for instance people displaced in the Hoa Binh hydropower dam. The resettlers have been receiving three assistance programs to mitigate the impacts of displacement, including the program 747 (between 1995 and 2001), 472 (from 2002 to 2006), 1588 (2009 and 2015), and the program 134 and 135. Although these programs have contributed to the improvement of livelihood, they haven't worked effectively and efficiently as expected. The investment scale is small and fragmented, and the expenditure used for agriculture restoration and development is not adequate. The agricultural program hasn't helped farmers produce competitive products in the market, so farmers do not benefit sufficiently. Payment for infrastructure has been enhanced, but there are about 40 percent communities lacking of schools, health care, and water supply systems (Nguyen Cong Quan, 2011).

3.5   Discussion and conclusion

The case of Vietnam allows us to draw a few conclusions about the current 'hype' around international land grabbing and to extend the debate to other fields that are no less relevant. Seen from the perspective of Vietnam, the global land grab debate consists of the emphasis on the role of foreign actors (foreignisation) in land acquisitions, and on the singular attention paid to acquisition of land for agricultural production purposes. Our point is not to deny the importance of these issues; rather, it is to assert that they are only part of more encompassing processes of change. The key problems arising from land grabbing—displacement and dispossession of resident populations in addition to sustainability issues—are not confined to the land-rich and people-poor peripheries of Africa and other areas where large land holdings are converted to corporate agriculture (or held for financial speculation). They are equally features of the structural transformations experienced by Vietnam and similar countries where an essentially rural economy and society is giving way to urban, industrial, and services-based equivalents. This is not an equitable process, but entails the transfer of resources (land, water, nature) in favour of stronger actors (developers, investors, state agents, etc.) at the expense of weaker ones (villagers, small-scale farmers, minority populations).

Land conversion procedures in Vietnam comply with the principle of free, prior and informed consent (FPIC) only in part. The Land Law (2003) prescribes timely announcement of a decision to convert ('recover') land, but the announcement often comes late—in practice, often just before project implementation. Compensation prices are set by local authorities; dissatisfied land holders may appeal but the decision is with the authorities, since the state is the ultimate owner of the land. In the case of private projects, compensation is in principle negotiated between interested parties; but since agreement is often difficult in many cases, the authorities intervene 'in the national interest'.

Large-scale land acquisitions are often interpreted as evidence of the rise in importance of 'vertical' forces over 'horizontal', territorial ones. Typically, global market forces (prices, opportunities for gain) are thus seen to overwhelm the regulatory powers of 'traditional' territorially based governments. While this case study of Vietnam actually confirms the role of globalisation as the harbinger of change, it also serves to nuance the vertical-over-horizontal discourse. Governments, horizontal actors *par excellence* and as such sometimes considered as creatures of the past, may very well actively implement the globalisation strategies. As shown in urbanisation and hydropower projects in Vietnam, the public sector assumes a leading role in bringing about these changes and reallocating land resources in favour of uses and users it deems more attractive. This is partly a logical consequence of Vietnam's socialist orientation, but it is not fundamentally different from cases observed elsewhere—for example, by Saskia Sassen (2005).

Land conversion is not a new phenomenon in development processes, but its extent and effects are. The process, on the one hand, could create massive benefits for stakeholders, such as government agencies, developers and investors, and local people. On the other hand, without a firm political commitment to protect the weak, land conversion also puts intense pressures on the livelihoods of farmers and others who have no stake in the rising industries. It is not the changes as such that need to be harmful, but the way in which they are played out. And that, indeed, requires a strong and committed 'horizontal' regulator.

References

Báo Quân đội Nhân dân. (2009). *Mỗi năm diện tích trồng lúa của Việt Nam bị thu hẹp 59.000ha.* Retrieved from http://www.baomoi.com/Home/DauTu-QuyHoach/ www.qdnd.vn/Moi-nam-dien-tich-trong-lua-cua-Viet-Nam-bi-thu-hep-59000ha/3 110840.epi.

Bộ Xây dựng. (1999). *Định hướng quy hoạch tổng thể phát triển đô thị Việt Nam đến năm 2020,* Hà Nội: Nhà xuất bản Xây dựng.

CODE, The Center for Development Consultation. (2010*). Study of the situation and effects of thuong kon tum on socio-economic and environmental factors - the proposal of displacement and resettlement consultation for the thuong kon tum hydropower dam.* Ha Noi: CODE.

Consultation Company on Electricity I. (2007). Result of livelihood surveys in the reservoirs of hydropower dam and resettlement planning for lai chau hydropower dam, Ha Noi: The Vietnam Electricity Cooperate.

Coulthart, A., Nguyễn, Q., & Sharpe, H. J. (2006). *Urban development strategy: meeting the challenges of rapid urbanization and the transition to a market oriented economy.* World Bank in Vietnam.

Department of Co-operatives and rural development. (2007). Displacement and resettlement policies for national programs in the mountainous areas and ethnic minority groups – problems and needs to be addressed, Ha Noi: Ministry of Agriculture and Rural Development.

DiGregorio, M. (2011). Into the land rush: facing the urban transition in Hanoi's western suburbs. *International Development Planning Review, 33*(3), 293-319.

Duy Huu. (2008). 'Deal with the jobs for land losers'. Available at http://www.baomoi.com /Giai-quyet-viec-lam-cho-nong-dan-bi-thu-hoi-dat/147/3220550.epi.

Firman, T. (1997). Land conversion and urban development in the northern region of West Java, Indonesia. *Urban Studies, 34*(7), 1027-1046.

GSO. (1996, 2001, 2010, and 2011). Ha Noi: Statistical Yearbook of Vietnam, Statistical Publishing House

Ho Moong Commune People's Committee. (2009). The situation of displacement and resettlement in the Pleikrong hydropower dam in Ho Moong Commune, Kon Tum Province: Ho Moong Commune People's Committee.

Hồng Minh. (2005). Hà Nội giải quyết việc làm cho lao động khu vực chuyển đổi mục đích sử dụng đất. *Lao động và Xã hội, số 270.*

Hung, N. T. (2010). *Urban air quality modelling and management in Hanoi, Vietnam* (Doctoral dissertation, Aarhus UniversitetAarhus University,[Enhedsstruktur før 1.7. 2011] Aarhus University, Danmarks MiljøundersøgelserNational Environmental Research Institute, Afdeling for Atmosfærisk MiljøDepartment of Atmospheric Environment).

Institute of Architecture and Planning. (2011). Economic Zones in Vietnam. Available at http://www.tinkinhte.com/cong-nghiep/khu-cong-nghiep-khu-kinh-te/toan-quoc-co-276-khu-cong-nghiep-va-khu-kinh-te.nd5-dt.135923.136147.html.

Institute of Development. (2010). *Displacement, resettlement, living stability and environmental and resources protection in hydropower dam projects (Research report)*. Ha Noi: Institute of development–CODE, Vietnam.

Institute of Strategy and Policy on Resources and Environment. (2009). *The status of resettlement programs in the hydropower dams in Vietnam.* Ha Noi: Ministry of Natural Resources and Environment.

Labbé, D., & Boudreau, J. A. (2011). Understanding the causes of urban fragmentation in Hanoi: the case of new urban areas.*International Development Planning Review, 33*(3), 273-291.

Labour News. (2012). The job training for rural labors is not adequate and low efficiency. Available at http://laodong.com.vn/Xa-hoi/Dao-tao-nghe-cho-lao-dong-nong-thon-ket-qua-rat-thap/82650.bld

Land Law of Viernam .(1993). *Luật Đất Đai.* Hà Nội: Nhà xuất bản Chính trị Quốc gia.

Land Law of Vietnam .(2003). *Luật Đất Đai.* Hà Nội: Nhà xuất bản Chính trị Quốc gia.

Lê Du Phong. (2007). *Thu nhập, đời sống, việc làm của người có đất bị thu hồi để xây dựng khu công nghiệp, khu đô thị, kết cấu hạ tầng kinh tế-xã hội và các công trình công cộng phục vụ lợi ích quốc gia.* Nhà xuất bản Chính trị Quốc gia. Hà Nội.

Luttrell, C. (2001). Institutional change and natural resource use in coastal Vietnam. *GeoJournal, 55*(2-4), 529-540.

Mai Thành. (2009). Về chuyển dịch cơ cấu lao động nông thôn sau thu hồi đất. *Tạp chí Cộng Sản* 15 (183).

Management Board of the Projects 747 and 472. (2009). The summary report of the living stability assistance for displaced people in the Hoa Binh hydropower dam, Hoa Binh: The Management Board of the Projects 747 and 472.

MOIT, Ministry of Industry and Trade. (2011). Strategic Environmental Assessment for the Electricity Plan between 2011 and 2015 towards 2030. Energy Institute: Ha Noi, Vietnam.

MOIT, Ministry of Industry and Trade. (2013). *Result of hydropower dam construction survey nationwide.* Ministry of Industry and Trade: Ha Noi, Vietnam.

Nghị Định số 69/2009/ND-CP (của Thủ Tướng Chính phủ) *về bổ sung quy hoạch sử dụng đất, giá đất, thu hồi đất, bồi thường, hỗ trợ và tái định cư.* Ngày 13 tháng 8 năm 2009

Ngo Viet Hung. (2007). The Changes of Land Use Plan and Impacts to the Poor in Vietnam Rural Areas. Paper submitted to the International conference on Sustainable Architectural design and urban planning, Hanoi Architectural University, Vietnam.

Nguyen Cong Quan. (2011). Lessons learnt for addressing the problems of resettlement in the Hoa Binh hydropower dam construction. Conference proceedings of Resettlement in the Hydopower Dam Construction: Do People have a better life? Ha Noi, pp.72-85.

NIAAP, Institute of Agriculture Planning and Design. (2008). The Master Planning of Resettlement in Ban Chat Hydropower Dam. Ha Noi: The Vietnam Electricity Cooperation.

Pawar, M. (2014). The Land Acquisition Act, 2013: Opportunities and Challenges ahead. http://landgovernance.org/proceeding/report-government-land-acquisition-india-china

Pham, X. N., Be, V. D., & Hainsworth, G. B. (2000). Rural development in Vietnam: the search for sustainable livelihoods. *Socioeconomic Renovation in Viet Nam: The Origin, Evolution, and Impact of Doi Moi. Ottawa: International Development Research Centre.*

Sàn giao dịch bất động sản An Biên. (2011). Việt Nam có bao nhiêu khu đô thị đạt tiêu chuẩn kiểu mẫu? Available at http://www.nhadatanbien.com/default.aspx?id=2803and pageid=newsdt.

Sassen, S. (2005). When national territory is home to the global: Old borders to novel borderings. *New Political Economy, 10*(4), 523-541.

Stockholm Environment Institute. (2007). Strategic environmental assessment of hydropower in Vietnam, 1993–2004, in the context of Power Development Plan VI. Ha Noi: The National Political Publisher.

Tan, R., Beckmann, V., van den Berg, L., & Qu, F. (2009). Governing farmland conversion: Comparing China with the Netherlands and Germany. *Land Use Policy, 26*(4), 961-974.

Tạp Chí Cộng Sản. (2011). Các khu công nghiệp ở Việt Nam: Hướng tới sự phát triển bền vững. Available at http://www.tapchicongsan.org.vn/Home/kinh-te-thi-truong-XHCN/2011/12504/Cac-khu-cong-nghiep-o-Viet-Nam-Huong-toi-su-phat-trien.aspx

The Tin Moi Newspaper. (2012). Hàng loạt khu đô thị mới, khu dân cư bị bỏ hoang.

Tran, T. V. (2008). Research on the effect of urban expansion on agricultural land in Ho Chi Minh City by using remote sensing method.

Tria Kerkvliet, B. J. (2006). Agricultural land in Vietnam: Markets tempered by family, community and socialist practices. *Journal of Agrarian Change, 6*(3), 285-305.

Van Suu, N. (2009). Agricultural land conversion and its effects on farmers in contemporary Vietnam. *Focaal, 2009*(54), 106-113.

van Westen, A. G. (2011). Land in China: Struggle and reform. *Development, 54*(1), 55-58.

Waibel, M. (2006). The production of urban space in Vietnam's metropolis in the course of transition: Internationalization, polarization and newly emerging lifestyles in Vietnamese society. *Trialog, 89*, 37.

Zoomers, A. (2010). Globalisation and the foreignisation of space: seven processes driving the current global land grab. *The Journal of Peasant Studies, 37*(2), 429-447.

4 COMPENSATION AND RESETTLEMENT POLICIES AFTER COMPULSORY LAND ACQUISITION FOR HYDROPOWER DEVELOPMENT: POLICY AND PRACTICE[13]

4.1 Introduction

There are increasing global concerns surrounding compulsory land acquisition in the public interest (Viitanen & Kakulu, 2008). Acquisition may inflict many adverse impacts on populations whose lands are expropriated (Alemu, 2013; Han & Vu, 2008; Kusiluka, et al., 2011; Syagga and Olima, 1996) including loss of income and job opportunities (farm and non-farm jobs), a loss of livelihood assets (land, common pool resources), as well as access to public services, and the breaking-down of social networks (Cernea, 1997; Ding, 2007; Jayewardene, 2008; Larbi, et al., 2004; Zaman, 1996). Land acquisition indirectly produces effects related to wealth redistribution; as farmers receive different levels of compensation, severe tensions arise between governments and farmers that burden the implementation of land policy and planning (Barrows & Roth, 1990; Hui & Bao, 2013; Zaman, 1996). The indirect impacts of compulsory land acquisition also may be substantial (Barrows & Roth, 1990; Zaman, 1996). In some cases, food insecurity is a serious problem arising from compulsory land acquisition (Cernea & Nogami, 2002). Social injustice arising from land acquisition is primarily related to inconsistencies in compensation policies in both horizontal and vertical dimensions. The former refers to variations that exist in the type and amount of land loss compensation received between different affected people whereas the latter implies differences in compensation types and amounts over time (Ding, 2004). In spite of formal protestations, most forcibly displaced people are left poorer than before displacement. As a result, the term "sustainable development"—so often used to justify forced eviction—may be challenged. If the government of Vietnam aims to increase energy security and improve livelihoods through eviction, it is crucial to explore the causes of this paradox and to search for solutions that turn forced displacement into smart development opportunities (Turton, 2009).

The legal framework for land acquisition of Vietnam has seen significant improvements since 1993 (Bui & Schreinemachers, 2011; Bui et al., 2013; Coit, 1998; Dao, 2010; Kim, 2004; Quang, 2002; Thu & Perera, 2011). From a country that did not have a land market in the 1980s, Vietnam shifted to a market-oriented land system. The most significant change was introduced through the 2003 Land Law that

---

[13] *This chapter was published as* Ty, P. H., Van Westen, A. C. M., & Zoomers, A. (2013). Compensation and Resettlement Policies after Compulsory Land Acquisition for Hydropower Development in Vietnam: Policy and Practice. *Land, 2*(4), 678-704.

granted land users more rights, especially the right to compensation. More importantly, decision 34/2010/QD-TTg of the Prime Minister in 2010 laid out the foundations for compensation, support, and resettlement with respect to irrigation and hydropower projects; herein the criteria for compensation, support, and resettlement schemes were rigorously elaborated. However, land has become one of the most important issues in recent years; Vietnam is not an exception in the global context of compulsory land acquisitions. The issue is the subject of increasing debate across the country in the run up to the new land law, which will be issued by the National Assembly of Vietnam in 2013. The foremost reason for contention is inadequate compensation and poor resettlement planning that generates hundreds of thousands of grievances annually and in turn reduces the trust of citizens in the political system of Vietnam. From 2008 to 2011, central and local authorities together received about 1.57 million grievances of which some 42 percent was addressed. Recently, the resistance of evicted people has become fiercer. Land loss protesters dress in red, wave flags, and unfurl banners at Government offices in Hanoi and Ho Chi Minh City, especially during times of important political events (The government inspectorate of Vietnam, 2012).

Land acquisition for hydropower dam construction is a good example to showcase these issues. Hydropower is viewed as an effective means to increase national energy security and thus within the public interest that can make use of the compulsory eviction mechanism. The recent survey by MOIT (2013) shows that there are 1,237 hydropower projects; within this figure, 899 are large-scale hydropower dams generating 24,888 MW of electricity. To date, 260 projects are operational, 211 plants are under construction to operate by 2017, and the rest is being licensed and registered. Additionally, there are 452 small-scale hydroelectricity plants either operating or under construction across the country. The construction of hydropower dams has displaced 44,557 households or about 200,000 people (Bui & Schreinemachers, 2011) and expropriated 133,930 hectares of land. Although hydropower dams have the potential to bring benefits in terms of power supply, flood control, and irrigation, their construction can harm ecosystems (Han & Vu, 2008; MOIT, 2013) and uproot local communities (Lerer & Scudder, 1999; Scodanibbio & Mañez, 2005; Scudder, 1997a; Scudder, 1997). In fact, Scudder (2001) claimed that large dam construction continues to cause impoverishment for resettled communities as well as and negatively affect millions of people living downstream. Also, a majority of studies revealed that the substantial losses in paddy land, sloping land, forests, grass fields, and water spaces lead to significant declines in farm outputs, as well as on-farm and off-farm income (Bui & Schreinemachers, 2011; Bui et al., 2013; CODE, 2010; Consultation Company on electricity I, 2007; Dao, 2010; Department of Co-operatives and rural development, 2007; Institute of strategy and policy on resources and environment, 2009). A salient feature of these displacements is that some 90 percent of the affected people in Vietnam belong to minority ethnic groups living in mountainous areas (CODE, 2010). Although the MOIT survey in 2013 does not explicitly mention the number of grievances caused by hydropower projects, it reveals that there is a significant increase every year.

The question is that the State takes highly consideration on improving legal framework but these policies cannot put into practice to ensure the equitable development among the State, developers, and displaced persons in the case of hydropower dam development. Therefore, the objectives of this paper are to investigate issues generated by hydropower dam construction and displacement, or more specifically by resettlement and compensation schemes associated with hydro dams and to examine the root causes of these problems.

Following this introduction, we review important theories and concepts formulated to analyse the issues surrounding the compensation and resettlement plan. After that, we discuss the legal framework of land acquisition and the gap between policy and praxis. Then, we introduce about the case study on the A Luoi dam and the displacement it has caused, in which we elaborate the mechanisms of land acquisition, compensation, and resettlement and summarise the problems that have been voiced against the compensation and resettlement process. The causes of these problems are presented in the subsequent section. Finally, following empirical evidence, we discuss the root causes of these issues and their implications for resettled people and then draw conclusions from the discussion.

4.2   Analytical Framework

Following negative experiences in the past, when some of its large-scale dam construction projects sometimes had negative social and environmental consequences, the World Bank has emerged as an authority in careful design of planning procedures. The revised involuntary resettlement objectives of the World Bank (2013) state that "displaced persons should be assisted in their efforts to improve their livelihoods and standards of living or at least to restore them, in real terms, to pre-displacement levels or to levels prevailing prior to the beginning of project implementation, whichever is higher". When compulsory displacement becomes inescapable, resettlement objectives must include minimised social risks and shocks, damage, and suffering; the protection of resettled people's well-being and rights; a facilitation of their rehabilitation among new hosts; and support for redevelopment and improved livelihoods at arrival sites (Sebenius, et al., 2005). To avoid impoverishment, good policy, proper resettlement planning and adequate resource allocation are critical (Cernea, 1997; McDonald-Wilmsen & Webber, 2010; World Bank, 2011; Yuefang & Steil, 2003). The principle of Free, Prior, and Informed Consent (FPIC) as a precondition for resettlement is just one step in developing a level playing field between local communities and government-sponsored, large-scale development projects. World Bank policy (OP 4.12) requires the application of the principle of FPIC in the case of indigenous communities. However, rights of members of majority populations should also receive reasonable protection in law and policy frameworks World Bank (2013). The World Bank (2004) also specified that the level of participation in a resettlement programme must be elevated to encompass collaboration and involvement in decision making instead of merely consultation and involvement in the execution of plans. In such cases, affected people are able to join a resettlement committee and participate in decision making or in designing resettlement programmes. Moreover, participation requires the involvement of multiple actors and organisations from local to national levels (Dao, 2010; McDonald-Wilmsen & Webber, 2010). On the side of the project developer, participation may help to avoid unnecessary and costly development (World Bank, 2004). We believe that participation of affected people in arranging resettlement programmes helps diminish adverse effects and severe vulnerability and thus enhances the chances of success for displaced people to adapt to new places. Furthermore, dams for hydroelectric power must be designed to maximise environmental and social benefits (World Bank, 2011) and ensure the rights and entitlements of indigenous people and ethnic minorities (WCD, 2000). In addition, development projects should respect existing land and access rights including those pertaining to customary and common properties. Further requirements include transparency in the negotiation process, fair profit sharing amongst affected people, project developers, and local government, as well as ensuring environmental sustainability, guaranteeing livelihood restoration, and respect of the local land policies (Zoomers, 2010). In addition, the World Commission on Dams (2000) emphasised that good governance makes for common ground between stakeholders in the negotiation and decision-making process. This includes equity, efficiency, participatory decision making, sustainability and accountability as key characteristics, in line with the concept of FPIC. In the case of investment of private sectors, the Performance Standard 5 of the International Finance Corporation (IFC) also emphasizes that investors should negotiate settlements with landowners rather than depending on compulsory land acquisition mechanisms of the government. If compulsory acquisition is unavoidable, private companies should prepare a supplemental resettlement plan besides the one of responsible government agencies to fully address the relevant Performance Standard (The International Finance Corporation, 2013).

Such international policies with respect to involuntary resettlement have been introduced to Vietnam through loans and investment by the World Bank, the Asian Development Bank and the work of international organizations such as the World

Dam Commission, and the International River Network. They have contributed significantly to the improvement of Vietnam's resettlement policies, especially for cases of compulsory land acquisition for hydropower dam construction (CODE, 2010; Dao, 2010). As a result, the Prime Minister in 2010 issued Decision 34/2010/QD-TTg in order to develop a clearer framework for compensation, support and resettlement policies. These aim to ensure that displaced people have places of settlement, can build a stable life with opportunities to develop production, raise incomes and incrementally improve infrastructure. The framework is intended to guarantee that they have better lives materially and spiritually in the long-term than before displacement as well as to emphasize harmonizing the interests of resettled people and already established inhabitants. The objective is quite similar to the World Bank's. The framework also states that the implementation process of compensation and resettlement must ensure democracy, publicity, fairness, transparency, proper purposes, proper persons and effectiveness. It appears similar to the FPIC principle; however, it is not concretized into specific guidelines for successive phases of land acquisition, such as planning, land recovery, compensation, displacement, resettlement, and livelihood restoration. The compulsory acquisition process implies that all activities of compensation and resettlement are designed and carried out by the responsible agencies of local government and then presented to affected people with details on losses, compensation values, and resettlement plan. There is no policy to put the people to be displaced in the centre of decision making process. For example, although article 21 of Decision 34/2010/QD-TTg declares that affected persons have the right to inspect and supervise the implementation of compensation assistance, and resettlement scheme; local authorities as a rule do not include any members from displaced communities in the committees preparing compensation and resettlement at the district level. All members tend to be recruited from project developers and functional departments of local councils. There is thus a big gap between policy and practice of compensation and resettlement policy in Vietnam (Dao, 2010; Han & Vu, 2008; Thu & Perera, 2011). Since the majority of displaced people for hydropower dam construction belong to ethnic minorities (Scudder, 2001), it would seem crucial to concretize the principle of FPIC in the planning and implementation process. Without this, the objectives of decision 34/2010/QD-TTg cannot be attained. Moreover, implementation and consultation should be undertaken by independent parties (e.g., local NGOs) to prevent local authority bias in favour of investors, as acknowledged in note 4 of Annex A- OP 4.12 by the World Bank (2013).

When it comes to compensation for loss, it must not only be just or equitable, but also effective in benefiting the landowners (Knetsch, 1983). That is to say that compensation in cash or land may not be sufficient to ensure that displaced people can restore and improve their livelihoods in the long term. This may require additional assistance such as training, *etc.*

Compensation for land is often complicated, particularly the estimation of land values. The *market value* is one option used. This is commonly defined as "the estimated amount that the land might be expected to realise if sold in the open market at valuation date after proper marketing between a willing seller and a willing buyer and they had acted knowledgeably, prudently, and willingly" (Asian Development Bank, 2007; International Valuation Standards Council, 2011). *Fair market value* might be used exchangeable with market value, but there is a distinction between them. The fairness of market value herein reflects the estimated price for the transfer of a property between willing parties who have the respective interests of those parties. It is necessary to carry out the assessment of the price that is fair for those parties taking consideration on the respective advantages and disadvantages that each is able to obtain from the transaction. Meanwhile, market value entails the strong points that are not available to market participants generally to be ignored, and therefore the concept of market value is narrower than fair market value (International Valuation Standards Council, 2011). The International Valuation Standards (2011) also differentiates between price and value. Price is the amount asked, offered or paid for an asset, value reflects the opinion that the most probable price to be paid for an asset in an exchange or the economic benefits of owning an asset. Because of financial capabilities, motivations and special interest of a given buyer or seller, the price might not reflect the true underlying value. As a result, market price is quite distinct from market value; they are equal when the market must provide sufficient information, efficient marketing, and prevailingly rational expectations. We can understand that the market price implies the negotiable capability of market value between market participants. In case of involuntary land acquisition, the government alone or in alliance with investors are willing buyers, but the affected landowners are often not willing sellers. As pointed out by (Miceli & Segerson, 2007) that the compensation paid to owners by using market price, whose land is taken, is systematically less than the amount owners would ask for their land in a consensual transaction because acquired land owners always response to compensation value by their subjective value or reservation price that reflects the market value. According to this understanding, compensation at market value often under-compensates unwilling sellers. However, it is very difficult to know the owner's subjective reservation price because self-interest induces owners to quote highly inflated values (Niemann & Shapiro, 2008). In addition, even landowners themselves may not know at what price they are prepared to sell. Another option is compensation at *replacement cost*. The replacement cost is equal to market value when the information about market value is reliable and comparable assets or acceptable substitutes are available for purchase. In most developing countries, however, conditions are insufficient to estimate market value and replacement cost, especially in remote and rural environment because the information on land prices is not reliable (Asian Development Bank, 2007).

Vietnam is an example. Decree 123/2007/NĐ-CP defines the market price as "the actual market price of land use right transfer as results of common actually completed transactions between transferors and transferees under normal commercial conditions and without the influence of factors that cause sudden irrational price rise or decrease". It also emphasizes that compensation price is valuated at market price. However, this is not put into practice when the government, especially the Provincial People's Committee (PPC), often applies improper methods to valuate land by collecting prices from offices of land registration, notary, and tax records to construct a land price framework on a yearly basis. As a result, land prices in the compensation framework is often far lower than market price (Thu & Perera, 2011; World Bank, 2011). The study of World Bank (World Bank, 2011) confirmed that more than 80 percent of resettled people in Vietnam are dissatisfied with compensation since land prices were much lower than market prices. Such differences in land prices between different valuation frameworks existing side by side lead to conflicts over land acquisitions throughout the country, especially on compensation rates (see Ty et al., 2014). The question is why provinces still retain the administrative pricing framework that is the cause of dissatisfaction. The first reason is that local governments want to boost capital investment by the private sector, especially Foreign Development Investment (FDI). Secondly, the government is also a major project developer, and therefore has an interest in modest prices for compensation costs (Han & Vu, 2008). The local government subsequently sells land at market prices through a competitive auction setting. Therefore, land investment often results in considerable revenues for provinces. For example, the revenue from land development, taxes and registration fees accounted for 15 percent of the total budget as reported by Thua Thien Hue province in 2012 (Thua Thien Hue People's Commitee, 2013) and more than 42 percent between 2003 and 2007 in Da Nang city (The Sai Gon Giai Phong, 2013).

In the case of hydropower development, land acquisitions are mainly located in the mountainous rural areas of the interior where the market price of land is not clearly defined because transactions are relatively rare (Marsh et al., 2007). The majority of tenancy contracts commonly take place between relatives; the social aspects of tenancy play a clear role in that tenancy contracts sometimes involve no rental payment (Phan & Fujimoto, 2012). Therefore, the compensation price is decided by Board of Compensation, Support, and Resettlement (BCSR) based on the provincial price framework which results in lower than market prices. In particular, compensation prices and payment process are decided by the compensation committee at district and provincial level without negotiations with people losing land

73

(CODE, 2010; Dao, 2010). For the time being, hydropower development in Vietnam is considered as a security purpose, not a commercial one according to the 2003 Land Law. As a result, the acquisition of land proceeds according to the compulsory mechanism. Besides this, there is also a voluntary land acquisition mechanism regulated in the 2003 Land Law. However, according to World Bank (2011), many private investors followed this voluntary procedure when investing in urban areas as they could thus avoid complicated administrative procedures with authorities at all levels. Conversely, voluntary procedures have not always been successful when rights holders of a small part of the needed land hold out in order to gain a windfall profit. The 2003 Land Law has no mechanism to resolve such situation (World Bank, 2011).

4.3   Research Design and Methodology

4.3.1   Research Problem

In view of the problematique discussed above, this study addresses the issue of a dam-related resettlement project near Hue city in Central Vietnam: How can the process of compulsory land acquisition for hydropower dam construction be described and understood? What problems are voiced by displaced people with respect to compensation and resettlement? and why? And what are the implications for improved policies towards the rebuilding of sustainable livelihoods and ensuring the wellbeing of forcibly displaced people?

4.3.2   Data Collection and Analysis

This research project used mixed methods to gain insight into the process of land acquisition for hydropower dam construction. It collected secondary as well as primary data for analysis. Therefore, both qualitative and quantitative methods were applied to collect both factual information such as information about household size, income sources and household information concerning the thoughts, ideas and experiences of resettled people. Both methods have strengths and weaknesses. By combining both methods we aim at a comprehensive approach.

Secondary data was collected at the district-level Department of Natural Resources and Environment (DoNRE) as well as at the hydropower company, including reports of the feasibility study, the environmental impact assessment, the land and property loss inventory, the land use survey, and descriptions of the displaced people. All legal documents on land recovery, compensation, and resettlement were downloaded from the online legal archive of Thua Thien Hue Documents.

Primary Data: The first interviews were conducted with district officials at the Department of Natural Resources and Environment and staff of the A Luoi

Hydropower Company who were members of BCSR and involved in most activities of compensation and resettlement plan. This yielded general information on the A Luoi hydropower dam, including the process of investment, compensation, and resettlement. We also gathered their opinions about problems that occurred when implementing the compensation and resettlement package, e.g., the number and nature of grievances, and how these have been addressed. Also field visits have been undertaken to the A Luoi dam construction site on the border between Vietnam and Laos, in a location that is not publicly accessible. We further interviewed the former headman to get the historical A Den village to collect information on living conditions before and after resettlement, as well as their opinions about the compensation and resettlement package since the former village headman was involved in the BCSR activities. Following that, a focus group discussion was undertaken with 8 key persons, including the former village headman, three elders, two young people who are A Den households but also worked for the former commune as police and cadastral officers, the patriarch, and one leader of the host commune of the new resettlement. The focus group discussion reviewed the compensation and resettlement process that focused on the participation of A Den households in making decisions. This served to check the information collected at the district Department of Natural Resource and Environment (DoNRE) and other sources. The main problems met have also been investigated at this meeting that helped us focus the subsequent household interviews on the principal problems after resettlement of A Den households. All 60 households of the village) have been interviewed in 2013 to collect household-level data as well as their experiences and views towards the compensation and resettlement scheme. The questionnaires covered such issues as the size of agricultural land and changes of income and food security after resettlement. After the household survey, we selected two households to have in-depth interviews on their experiences of compensation and resettlement. Household data was coded and analysed in SPSS 20. After screening the data, the descriptive analysis tool was used to present the relationship between selected variables in custom tables. Data from the generated tables and graphs were interpreted to answer the research questions. Finally, we had an interview with a former Vice Minister of Ministry of Natural Resource and Environment (MONRE) in September 2013 in Ha Noi to discuss the issues of compensation and resettlement policy and practice.

Limitation of Research

The recall method to collect data on conditions before and after resettlement relies on the memory of participants. This does not ensure absolute accuracy of household data, especially on income that was difficult to estimate exactly for households having both cash and subsistence income. The process of resettlement could have been experienced as negative, which might have induced participants to idealize the situation prior to resettlement. Households often did not disclose how much money they have received in compensation and often said "we have spent all compensation money because it was little". We intend to undertake more research to add robustness to this issue, and answer other questions such as: who did benefit and managed to move upwards after resettlement? Who are most vulnerable after resettlement and move downward? Why? How do resettled people adapt to new settings? What are the roles of different stakeholders in mitigating and enhancing the results of resettlement policy?

4.4   Results and discussions

4.4.1   Study site and historical of displaced people

In 2007, construction of the A Luoi hydropower dam started on the A Sap River, a branch of Mekong river system located on the border with Laos. Owned by the Central Joint-Stock Hydropower Company, the dam became operational in 2012 and aims to produce 686 million kWh per year. The total area of land acquisition was over 2,080 hectares, of which over 95 % was crop and forest land. These lands were used by different types of landholders: communes managed 393 hectares, the State Protection Forest Management Boards handled over 54 hectares, and individual households used 1,633 hectares. The dam and reservoir displaced 218 households (about 872 villagers), mostly ethnic minorities.

One of the affected villages, A Den, was selected for an investigation into the compensation and resettlement process, the grievances related to compensation and resettlement, as well as the magnitude of changes and consequences resulting from the dam construction. A Den village, Hong Thai commune, was located about 300 m from the commune office and 7 km from A Luoi Township (see Figure 4-1). It was situated in the valley between tributaries of the A Sap and Ta Rin Rivers. The village was divided into five blocks; each block consisting of several houses separated by a small concrete road. Each house was surrounded by a garden, usually planted with a mixture of crops including cassava, corn, coffee, vegetables and fruit trees. A barn was usually located near the house and was used to keep cows or buffaloes.

Besides the garden, wet paddy fields and fish ponds were located along the stream. Other agricultural land and forest tree plantations were situated at some distance from the houses. The total residential area of the village, 32 hectares, was inhabited by 61 families or around 274 people of the Ta Oi ethnic group. The Ta Oi originated from Laos where they practiced slash and burn farming and used open areas for planting dry land paddy in addition to relying on the natural forest for food, medicines and raw materials. In 1972, Ta Oi people started moving from Laos to Vietnam to settle in Hong Thai commune, A Luoi district. They named their village A

76

Den, which means "blissfully great harvest" in the Ta Oi language. Nonetheless, the move to Vietnam did not contribute greatly to their living conditions. However, livelihood conditions gradually improved from 1993 onwards when the Vietnamese government issued the land law and launched policies that targeted ethnic minorities. People were given more freedom to develop farmlands as access to land and markets improved. Also, access to education, health services, infrastructure and electricity were improved, as the village was located close to the administrative centre of Hong Thai commune.

Table 4-1 Information of respondents represented to interviewed households (*n* = 60). *Source: Authors' household survey, 2013*

		# of Respondents	%
Ethnicity of interviewees	Ta Oi	58	96.7
	Kinh	2	3.3
Gender of interviewees	Male	24	40.0
	Female	36	60.0
Education level of interviewees	Illiterate	20	33.3
	Primary	14	23.3
	Secondary	19	31.7
	High school	7	11.7
Occupation of interviewees	Unemployed	9	15
	Farmer	42	70
	Carpenter	2	3.3
	Government official	7	11.7
Economic status of households	Non-poor	34	56.7
	Poor	26	43.3

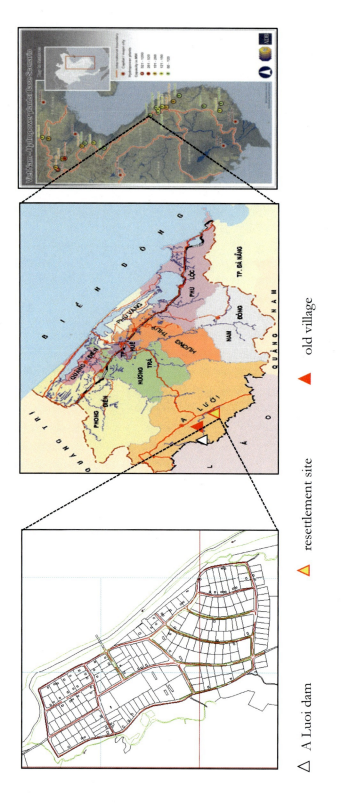

$\triangle$ A Luoi dam $\blacktriangle$ resettlement site $\blacktriangle$ old village

Figure 4-1 Map of the Can Tom 2 resettlement village, A Luoi district.

Source: Institute of Energy, 2011; Thua Thien Hue PPC, 2005

78

In 2007, residents of A Den village received an announcement stating that the A Luoi hydropower plant would be constructed and the whole village would be evicted. Financial compensation for their loss of assets was given on 13 May 2010 and people were informed to move to the resettlement area or elsewhere in June 2011. Since November 2011, 61 households from A Den have been living in the relocation village named Can Tom 2. Located in Hong Thai commune, the resettlement site it is about 15 km from the old village. Table 4-1 shows some of the characteristics of the Ta Oi households. Their education level is quite low; two-thirds are farmers and nearly half of the residents are classified as poor. Observations also revealed that most labourers working in agriculture and forestry are young people who have obtained either primary or secondary education.

4.4.2   The Investment and Land Acquisition Process

The Prime Minister approved the proposal for the A Luoi hydropower project in 2005. The project was supported by the Ministry of Industry and Trade under the BOO (Build, Own, Operate) investment model. The investment capital was guaranteed by the Ministry of Finance via foreign loans to import facilities. The Development Support Foundation was responsible for implementing the displacement and resettlement programme as well as for producing the domestic facilities in support of the project. The Prime Minister's decision also secured the sale of electricity to the Electricity Cooperate of Vietnam (or EVN) and facilitated Thua Thien Hue Provincial People's Committee's (PPC) project approval. In 2008, and because the investment was also consistent with the planning of hydropower and electricity development adopted by the Ministry of Industry and Trade (MOIT), Thua Thien Hue province granted the investment certificate to the Central Hydropower Join-Stock Company. The investor receives many incentives from the province. For example, the project does not have to pay income tax for 4 years from the start of dam operations; after this period, the investor only has to pay half of the regular tax for the next 9 years. Additionally, while other businesses must pay the regular income tax of 25 percent, the hydropower company pays only 10 percent income taxes. The company also receives exemption from the land lease tax for 15 years as well as import taxes. Such financial incentives have prompted several private companies to look for hydropower investment opportunities in the Central provinces. After all, projects such as the A Luoi hydropower dam are lucrative. The dam has generated electricity since June 2012 through two power stations; the company expects to reach revenues of over 300 billion VND (or roughly US$14.4 million) and a pre-VAT profit of 25.7 billion VND (approximately US$1.2 million) in 2012 (The Central hydropower join-stock company, 2011). Vietnam is facing a shortage of electricity and has to

79

import power from China. Since demand is increasing, profit seems to be secured over a longer term.

After the PPC approved construction of the dam in 2008, the Department of Natural Resources and Environment (DoNRE) started collecting land use and cadastral maps of the affected area to delineate areas which would be flooded as well as to make land allocation maps for the hydropower company. Once the PPC approved the acquisition and allocation of land to the investor, information about those decisions were posted in the communes and villages. At that time, these notices served as the first information affected people received about their land loss. Households did not have any role within this process since the project was considered to be a technical issue to be decided by the province and district with the assistance of DoNRE. People losing land under the scheme did not have the right to refuse land acquisition decisions; the land law awarded the government the right to take land back for reasons of public interest, national security and economic development. Affected people simply had to accept this regulation (The National Assembly of Vietnam, 2003). According to the 2003 Land Law, they are left with grievance rights, including denunciation and complaints about the compensation packages. Complaints here refer to appeals to review administrative decisions and the like when complainants have grounds to believe that such decisions or acts contravene laws and infringe upon their legitimate rights and interests. Denunciations are made by displaced people who claim that their legitimate rights and interests have been damaged by authorities against the law (Author's interview with district officials, 2013).

Despite this established grievance process, it occurs after the compensation and resettlement procedure has already taken place. In the case of the A Luoi hydropower project, the investor first hired a consultation company to prepare a master plan for compensation, assistance, and resettlement for submission to PPC for approval. After acceptance, PPC established a management board to implement the ground clearance, compensation, support and resettlement (BCSR) process at the district level. In this steering committee, leaders of the district, finance department, and the hydropower company were represented. Other members included representatives of line agencies of the district. DoNRE was responsible for mapping, land and property inventory, and land use certificate allocation. Commune leaders and representatives within the commune of the Fatherland Front Committee worked as a bridge between BCSR and affected households; they organised meetings with households to announce the details of land recovery, explained compensation and resettlement policy as well as persuaded people to accept the plan. Only one villager was invited to join BCSR; he was the village head who was also a district official. Next, BCSR photographed the properties of affected people to prevent households from planting more trees in order to claim more compensation. Right after recording the effects of the expropriated households were recorded, including houses, crops, forests, and other structures, BCSR sent the land appropriation decision to each household. After this, an asset inventory was conducted together with affected households, village heads, and commune cadastral officials. All information was used to assert the status of lands and assets in order to validate the rights of land users as eligible for compensation or not in terms of the provincial regulations. Then, BCSR decided upon the compensation prices for land, houses, crops, forest trees, and other assets according to the compensation pricing scheme of the province as well as the assistance package for each family. This information was sent to each family by commune officials from door to door and posted at the commune office in 2008. Payment then was distributed in several rounds for different affected households.

4.4.3 Compensation and Resettlement for A Den Households

The main principle of compensation was "land for land", *i.e.*, dispossessed land rights holders could receive an equal amount of land in the new settlement which in principle corresponded to their land loss. In cases where this was not possible, an additional payment would be provided to make up for the difference.

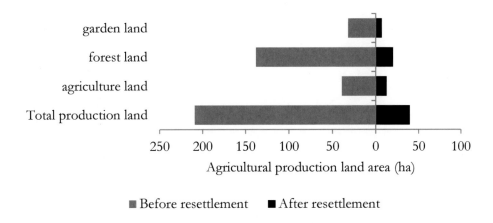

Figure 4-2 Production land of households in A Den village before and after resettlement (*n* = 60). *Source: Authors' household survey, 2013.*

Figure 4-2 shows that A Den households lost a considerable amount of land due to dam construction. The total area of the village was reduced by about 81 percent; each family lost around 2.8 hectares on average, from 3.5 to 0.7 hectares on average. The loss of land and land-based properties was compensated in cash in May 2010 with each family receiving between US$2,500 and US$15,000. In addition to the compensation, the investor supported displaced people with other types of assistance, including rice supplies for 24 months, electricity for one year, US$80 for displacement-related costs, a one-time provision of fertilizer for improving soil quality, several agriculture trainings and US$90 for job training, and several pigs, chickens, and ducks for livestock restoration. Furthermore, children of displaced families had free education for one year.

The district offered households a choice between two new places of residence after displacement in 2009. As the first option was located in a remote mountainous area, resettlers complained that they had to select the second option because it is closer to the centre of A Luoi Township, more convenient and accessible to the market and transportation, and on flat land (interview with former village headman). In reality, it is very difficult to find a suitable place for resettlement when free land is in short supply in the region. In November 2011, A Den households moved roughly 8 km from the old village to the new resettlement area named Can Tom 2 village in Hong Thuong commune where the investor constructed houses for the displaced households. Each family received a house surrounded by around 60m² of garden land. The resettlement site is equipped with concrete roads, kindergarten and primary schools, gravity irrigation and a drinking water supply, electricity, a health clinic, community houses, and playgrounds.

4.4.4 Problems after Resettlement

Many grievances were aired during interviews with resettlers. Not only did they express their criticism verbally, but they also sent official grievance letters to the relevant authorities to voice their opinions about compensation and resettlement. According to the district-level DONRE, about 90 percent of resettled households accepted the compensation packages. In 2008, the Commune received 110 letters complaining about low compensation prices while 84 letters were sent to the district to denounce the measurement of land and property. In 2009, a collective letter on behalf of the whole community was sent to the district authority and the hydropower company to request more support for livelihood restoration and development in the new settlement. In addition, seven households appealed against the inappropriate recompense for houses, crops, and land. In 2010, 60 households did not accept the compensation price because of disputes over land boundaries among households in the course of the land inventory process. In 2011, households sent another collective letter to ask for compensation for the loss of coffee lands (interview with district officials, 2013). A survey among 60 households of former A Den households showed that grievances occurred constantly. The reasons are discussed in the following section.

Inadequate compensation due to low pricing, insufficient and unfair payment: In general, Ta Oi people in A Den village were not satisfied with the compensation. In fact, nearly 86 percent of households complained that the compensation amount was much lower than the value of what they had lost; more than 80 percent disagreed with compensation prices since they were low compared to actual market price observed in the region (see more in Figure 4-3). Affected households also said that they were compelled to accept compensation payments because it was the provincial pricing scheme; if the payment was rejected, farmers would be left with nothing once their land was flooded. Moreover, about 10 percent of households, especially women, did not know the exact amount of compensation. According to one woman, "I didn't know about compensation my husband only told me that he received compensation money but spent all of it on extending the house and on petrol for transportation during displacement."

Figure 4-3 Self-evaluation of respondents on compensation scheme ($n = 60$ households). *Source: Authors' household survey, 2013.*

Others were also dissatisfied with their compensation. A 39-year-old male Ta Oi villager said that he lost 2.95 hectares, including a garden of 1,250 m² where he kept livestock and grew fruits, vegetables, cassava, and maize. He also lost 1 hectare of forest land planted with Acacia, 1,500m² of upland rice, reclaimed lands 15,000m² as well as coffee land measuring 1,200m². In the resettlement site, he received a land area of 12,80 m², including housing land of 200m², forest land of 10,000m², garden land of 1,600m², and a rice field measuring 1,000m². He did not get compensation for the land he contributed to the land pool for the coffee company. Therefore, he is strongly displeased with the compensation he received. Another 40-year-old Ta Oi male claimed that he lost 1.4 hectares of land. His garden, measuring 350m², was used to grow fruits, bamboo, and to raise livestock. He also held 50m² of residential land, and1 hectare of annual cropping land that he used to grow upland rice, cassava, maize, and coffee. He accompanied the company and district officials during his land survey and it was correct. He also went to 4 meetings with the commune authorities and company representative. However, they decided the compensation price of lost land, not him. Despite his previous assets, he received 1,800m² of garden land and 200m² of residential land in the relocation site. He did not receive compensation for the 1 hectare of forest land or for his coffee land. He said that he did not agree with the compensation price because it was much lower than the market price of land. Currently, he still has 8 ha of forest land in the old village, but it is located too far away to continue planting Acacia (Author's in-depth interview, 2013).

In the previous settlement, A Den households planted Acacia, upland rice, cassava, and maize on reclaimed land but only half of them were compensated for this loss. In addition, 44 households reclaimed and contributed 26 hectares of land to the Quang Tri coffee company land pool; they also worked as labourers for the company (household interview, 2013). They received income for their land contribution as well as for planting, monitoring, and harvesting coffee. Due to dam construction, this land was flooded but not compensated for. The hydropower company explained that as the coffee land was leased to the Quang Tri coffee company, they paid the coffee company for the loss of the coffee trees only. They were not expected to compensate households for this land, as this was not eligible under the provincial regulations. Those who lost land in this way sent a collective letter in 2011 to the commune, district, and to the investor to request cash compensation. In response, the commune requested the district and the hydropower company to estimate the value of the coffee land and to inform all affected households. In September 2011, the district confirmed that the value of compensation, around US$133,000, for this coffee land loss was correct. However, compensation for coffee land loss has not been paid to A Den households since 2011. Furthermore, the remaining land of displaced households, which is not being flooded by the dam, encountered with another problem when the coffee company went bankrupt in 2011 because they used the land lease certificate to acquire a loan from the Quang Tri Bank before that. Therefore, displaced households could not access to the remaining land not flooded anymore because land belonged to the bank. As a consequence, the district had a plan to buy the land back from the bank because coffee plantation was set out as an agricultural development priority strategy of the district and to keep stable security in the district (Author's interview with district officials, 2013).

Moreover, even though households could no longer access parcels of non-flooded land as it was blocked by the reservoir, farmers did not receive any compensation for this loss. More importantly, not all promised compensation money for flooded land and trees was paid out. Most resettled households only received half of the promised amount in cash; the rest that was to be compensated through the allocation of plots of comparable land was not given. Furthermore, the late payment of compensation package caused unfairness amongst land losers. This is because the prices for compensation were decided in 2007 after inventory, but payment was made in different years between 2008 and 2011. Meanwhile, every year the province adopted another price framework, often significantly increasing land prices. Also, market prices for crops like cassava had increased dramatically. Many of displaced households accepting payment in earlier years received less than later payments. As a result, resettled people who had received compensation in the earlier years were angry about what they perceived as unfair treatment. These farmers sent a collective grievance letter to the district and to the hydropower company to ask for compensation and for the change in pricing levels between 2008 and 2011. The province urged the company to accommodate these complaints so as to reduce the social and political tension in the region where several hydropower dams are still being constructed. Since the province provided the investor with many favorable conditions, it was felt they had to accept the decision of the province.

Low participation levels and lack of transparency in the compensation and resettlement process: As can be seen from Figure 4-4, some 55 per cent of displaced households stated that the project developer determined the process of compensation and resettlement, while 45 percent indicated that the commune and district government had directed the process. Only 5 percent of the households, those who can be considered as village leaders, confirmed that they participated in resettlement site selection, along with district officials and the project developer. These findings illustrate the limited involvement of displaced people in the selection process of the resettlement area. Although 95 per cent of households confirmed that information was transmitted through meetings in the commune office, households did not have a real bargaining position to negotiate their needs. The meetings served only for the dissemination of information after all decisions had been made by the province and district; displaced people could convey their opinion to the commune and district but it was not considered important at that stage. Participants said that they had voiced their opinions and needs in several meetings with the investor, district, and commune but that their views did not influence those decisions. The investor and authorities typically explained that all decisions were based on the policies of the Party and the State and implied that people must accept their decisions. Thus, the selection procedure of resettlement area was not a participatory process because it did not include the active involvement and consultation of the affected people. In fact, dissemination is a one-way transfer of information offering affected people no options for involvement in the decision making process (World Bank, 2004).

Low commitment: As mentioned by resettled people in the focus group discussion, the investor and the district organised meetings with them and made many promises. Among these were that affected households would receive similar houses and plots as in the former village. In addition, the resettlement site would have sufficient electricity and an adequate water supply for drinking and agriculture. Households would also be allocated between 1.5 and 2 hectares of land for farming and agro-forestry. Similarly, over 83 per cent of households expressed strong disappointment about unfulfilled promises; they complain of poor housing quality and lack of repair work by the investor. Moreover, many bomb-holes dating from the war surround their houses. As these have not been filled, they cannot grow fruit trees and vegetables in their garden plots. Furthermore, households have not received the promised farm land. Annoyed, most resettled people stated that, "We haven't seen the hydropower company since being resettled. They have disappeared already". By the 2003 Land Law, the hydropower company does not have any responsibility for displaced households after the completion of resettlement construction and payment. While resettled people face many challenges in the relocation area, the company considers their duties in terms of compensation, assistance, and resettlement as finished.

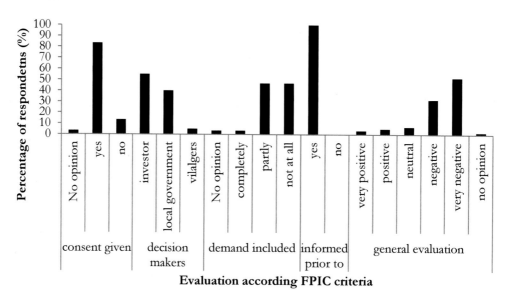

Figure 4-4 Respondents' evaluation of compensation and resettlement process according to FPIC principle. *Source: Authors' household survey, 2013.*

As a consequence, most resettled people are angry and so try to obtain more support from the hydropower company through the commune and district offices. Their needs cannot be met completely since the hydropower dam just started generating electricity and therefore the company cannot afford the costs; neither does the local government have the budgetary means to satisfy their demands. As a result, nearly 90 percent of affected households said that the construction of the hydropower dam has made their life worse than before.

Significant loss of income and deterioration of food security. Ta Oi resettlers report that before losing land to the hydropower project, each household could earn about VND 32 million (roughly US$1,600) a year. Of which, 59 percent of this income was sourced from crop production (cassava and rice); about 23 percent derived from livestock, 9 percent from government salaries, and 9 percent derived from wage labor, migration, and forestry and non-timber forest products (Household interview, 2011; see more in Table 4-2 about the distribution of income before and after resettlement). In contrast, currently each household only earns roughly VND 6 million annually; this is a reduction of 80 percent. The income from government jobs (only 7 households in the village) account for 85 percent of total village income, 11 percent is from wage labor, and 4 percent from livestock (Household interview, 2013). Figure 4-5 shows that there is a significant loss of income sources after resettlement. Currently, no household is able to generate income from agricultural production and forest. Households instead earn a very small income as day labourers on Acacia plantations in the region; this occupation is less-desirable as it is temporary and available only during the harvesting season. Poor households in particular have incurred income losses of up to 91 percent in comparison to the pre-resettlement situation.

Table 4-2 Household income distribution before and after resettlement. Unit: VND million (1 USD = VND 20 million). *Source: Authors' household survey, 2013*

	Sum	Mean	Max	Min	Range	Variance	Standard deviation
income before displacement	1,428	32	243	0	243	1,457	38
income after resettlement	263	6	75	0	75	215	15

Furthermore, as annual food production declined significantly after resettlement, food security is now a greater concern. The satisfaction of nutritional needs from subsistence food production dropped from 53.3 percent in the old village to 5 percent in the new settlement. Before resettlement all households produced food, with an annual average production of 5,798.7 kg per household, whereas now they produced only 93.1 kg per household annually. Not only do affected families produce less food for their own use, household expenditures on food have declined considerably. This implies that the resettled population has less money to spend on food than before. In the survey, 11 households have no income, in money or in-kind, and no food expenditures. These households are extremely vulnerable to the effects of food insecurity.

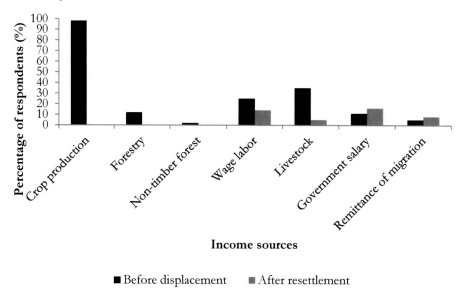

Figure 4-5 Accessible income sources of households before and after resettlement (*n* = 60). *Source: Authors' household survey, 2013*

Moreover, all resettled households complained that they could not produce subsistence crops because their compensation land is too small. In addition, the land is not cultivable because of poor soil quality. Since the Ta Oi depend much on farmland and forest, losing farmland means that they lose the most important source of income and food. As explained by district officials, it was very difficult to find sufficient land in the region to allocate to resettled households because host communities already occupied the surrounding land. Presently, the land of host households in Hong Thai commune as well as the protected lands belonging to the A Luoi Forest Management Board encircles the resettlement site. As a result, resettled people cannot find land to continue to cultivate through their traditional slash and burn practices (Author's interview with district officials, 2013).

Resettlement community disorder and lack of social cohesion: In addition to the 60 Ta Oi households from A Den village who account for 56 percent of the total population of the resettlement site, there are about 46 households resettled from other affected villages in five communes. Among these, Pa Co households comprise 36 percent and Kinh families make up 14 percent. As they live in separate areas within the same commune, the district decided to regroup them into Can Tom 2 village. However, as the Ta Oi and Pa Co do not want to live in the same village, each has requested separate villages. Each community had their own leaders and patriarchy in their previous villages who now want to maintain their role in their community. As the leaders of each community explained, Pa Co and Ta Oi are quite different ethnic groups who hail from different origins and are distinct in terms of culture and language. Nonetheless, they understand each other easily because percent of the similarities between their languages. Moreover, according to the survey with Ta Oi and Pa Co people, both groups speak the Kinh language fluently and their children study in the Kinh language in the same schools. At present, the district has not decided on the arrangement and name of the resettlement site, and they are awaiting directions from the province authority. The lack of a village administration board has caused many issues and there are no political, social, and professional associations such as a farmer's union or unions for women or youths. These unions are typically a part of the political system in Vietnam and thus are present at all administrative levels of central, province, district and commune. The farmers union especially is essential since it organises many activities for farmers, such as sharing cultivation experiences, trainings, and social activities (interview with host commune leaders, 2013). Although the district assigned a temporal village headman, he said that he had no role except for participating in meetings organised by the commune and district committee. Many resettled people said that they did not know who their village leaders are. Therefore, it was difficult to find the temporal village headman. Additionally, people in the focus group discussions disclosed that while there is a health clinic in the resettlement site, there is no doctor to operate it. When they are sick, they go to the clinic of Hong Thuong commune where they are refused access because their health cards are eligible only in the old commune. Consequently, they have to rely on the clinic in the old location, which is located at some distance and so is inconvenient. There are also reports of conflicts between resettled people and host households in Hong Thuong commune as they practice slash and burn farming techniques on lands belonging to the host population. In retaliation, the host population has broken their water supply pipes (Focus group discussion, 2013).

4.4.5   Discussions and Policy Implication

The implementation is mostly carried out by local authorities and the investor in this case study. In this alliance, local government tends to align with the investor as both stand to benefit from the project (Le, 2013; Thanh, 2013). In Vietnam, many researchers point out the weaknesses of compensation and resettlement procedures, such as low compensation prices, low participation and lack of long-term commitment to share benefits with affected people. However, the root cause of those limitations lies in the fact that government from the central to the local level tends to favor project developers rather than people losing their land. According to a former Vice-Minister, this type of coalition can be considered as an "interest group" (*nhóm lợi ích*) that is the biggest challenge for integrating sound principles for involuntary resettlement policies recommended by international organizations such as World Bank, ADB, and WCD, as well as for to the proper implementation of the good points of Vietnam's land acquisition policies (expert interview, 2013).

Policy Implications

The results show the urgent need to improve the national development policy on electricity production and land acquisition for hydropower dam construction in Vietnam. This might start from changing the investment mechanism for hydropower. Currently, electricity in Vietnam is mainly generated by the Vietnam Electricity Group (EVN), a State enterprise. Although many private companies are constructing hydropower dams, electricity generation is still a closed sector. Only Vietnamese companies can take part in producing and selling electricity, whereas foreign companies can participate in the construction of dams. Therefore, we suppose that electricity is not a free market-based activity. This situation lends itself quite well to a host of benefit-sharing mechanisms that also account for the interests of those who have had to surrender their livelihoods in order to make hydro-power possible in the first place. As it is now, the investors need not consider the full cost of the investment in hydropower dam construction as the cost for resettlement and compensation is not well-defined. To date, there are few studies that carefully determine the balance between cost and profit of constructing hydropower dams if the investment includes the benefit sharing. Therefore, there is an urgent requirement to resolve the fundamental economic problem of hydropower projects, *i.e.*, the accurate estimation of the real benefits and costs, including those of introducing suitable resettlement and benefit-sharing programs that also take the long term view. This is necessary to satisfy the legitimate concerns of all stakeholders, including government, developer, displaced people, and consumers. Only when a long-term benefit sharing mechanism is made compulsory for hydropower investment will displaced people be able to restore and improve their living conditions. To strengthen the effectiveness of the benefit sharing mechanism, the new Land Law under discussing must change the purpose of land acquisition for hydropower development from national security to that of a commercial goal, since all hydropower companies now invest for profit. All hydropower companies should be treated equally in electricity production and selling, including the responsibility to share profit, including state corporations such as EVN. The experiences of other countries, including China, Brazil, India and Japan, have shown several successful cases of benefit sharing models (Mathur, 2013), turning of the impoverishment of displaced people into an opportunity for development (Turton, 2009).

A successful approach to compensating displaced people would need to address three concerns. First, it should enable resettled households to create a livelihood within their abilities. Younger people, with sufficient education and skills training, may be able to move into non-farm activities. However, displaced people especially older people, do not have such opportunities and thus need land for farming and access to forest and other common pool resources. Second—and this actually emanates logically from the previous point—people need to be actively involved in deciding and planning for their life and living conditions after displacement. Only real participation as enclosed in FPIC can prevent the bitterness and frustration that now often characterise resettled communities. The FPIC principle is fully carried out if the role of community-based organizations and local NGOs is strengthened in the process of land acquisition, compensation, resettlement, and in the benefit sharing mechanism.

4.5   Conclusion

The case study found that the majority of households were disappointed with the compensation and resettlement scheme. Complaints often mention the lack of arable land, poor soil quality, food insecurity, loss of income and job, loss of access to forest, inadequate and unfair compensation, and the difficult resettlement site. The efforts of the State to improve policy by issuing better land laws and specific guidelines for compulsory land acquisition for hydropower dam construction are not sufficient to result in effective compensation and resettlement policies. The process of investment and land acquisition for the A Luoi hydropower dam shows that the local government is exercising a top-down approach to compulsory land acquisition by imposing the planning and construction decisions, the compensation prices for losses by affected people, and the resettlement site selection. The participation mechanisms in the compensation and resettlement process for people losing land as well as other local NGOs and community-based organizations are not well-defined. The involvement of affected people in decision making is not mandatory in compulsory land acquisitions. There is no room for negotiation as all decisions are pre-determined without transparency. There is very poor consultation, collaboration, and little choice open for affected people in the compensation and resettlement plan. Hundreds of hydropower dams are being constructed and planned in the coming years, and therefore good governance measures of land acquisition must be urgently put into practice.

References

Asian Development Bank. (2007). *Compensation and valuation in resettlement: Cambodia, ADB, People's republic of china and india.* The Asian Development Bank: The Philippines.

Barrows, R., & Roth, M. (1990). Land tenure and investment in african agriculture: Theory and evidence. *The Journal of Modern African Studies, 28*(2), 265-297. Retrieved from http://www.jstor.org/stable/160863

Belachew Yirsaw Alemu. (2013). Expropriation, valuation and compensation practice in Ethiopia. *Property Management, 31*(2), 132-158. doi:http://dx.doi.org/10.1108/02637471311309436

Bui, T. M. H., & Schreinemachers, P. (2011). Resettling farm households in northwestern Vietnam: Livelihood change and adaptation. *International Journal of Water Resources Development, 27*(4), 769-785. doi:10.1080/07900627.2011.593116

Bui, T. M. H., Schreinemachers, P., & Berger, T. (2013a). Hydropower development in vietnam: Involuntary resettlement and factors enabling rehabilitation. *Land use Policy, 31*(0), 536-544. doi:10.1016/j.landusepol.2012.08.015

Bui, T. M. H., Schreinemachers, P., & Berger, T. (2013b). Hydropower development in vietnam: Involuntary resettlement and factors enabling rehabilitation. *Land use Policy, 31*(0), 536-544. doi:http://dx.doi.org/10.1016/j.landusepol.2012.08.015

Cernea, M. (1997). The risks and reconstruction model for resettling displaced populations. *World Development, 25*(10), 1569-1587. Retrieved from http://search.proquest.com/docview/61524231?accountid=14772

Cernea, M. M., & Nogami, H. (2002). The economics of involuntary resettlement: Questions and challenges. *Developing Economies, XL*(1), 101-106. Retrieved from http://search.proquest.com/docview/39084280?accountid=14772

CODE. (2010). *Displacement, resettlement, living stability and environmental and resources protection in hydropower dam projects.* CODE: Ha Noi, Vietnam.

Coit, K. (1998). Housing policy and slum upgrading in ho-chi-minh city. *Habitat International*, *22*(3), 273-280. doi:http://dx.doi.org/10.1016/S0197-3975(98)00011-3

Consultation Company on electricity I. (2007). *Result of livelihood surveys in the reservoirs of hydropower dam and resettlement planning for lai chau hydropower dam*. The Vietnam Electricity Cooperate: Ha Noi.

Dao, N. (2010). Dam development in vietnam: The evolution of dam-induced resettlement policy. *Water Alternatives, 3*(2), 324-340.

Department of Co-operatives and rural development. (2007). *Displacement and resettlement policies for national programs in the mountainous areas and ethnic minority groups - problems needs to be addressed*. Ministry of agriculture and rural development: Ha Noi.

Ding, C. (2007). Policy and praxis of land acquisition in china. *Land use Policy, 24*(1), 1-13. doi:10.1016/j.landusepol.2005.09.002

Han, S. S., & Vu, K. T. (2008). Land acquisition in transitional hanoi, vietnam. *Urban Studies, 45*(5-6), 1097-1117. doi:http://dx.doi.org/10.1177/0042098008089855

Hui, E. C. M., & Bao, H. (2013). The logic behind conflicts in land acquisitions in contemporary china: A framework based upon game theory. *Land use Policy, 30*(1), 373-380. doi:10.1016/j.landusepol.2012.04.001

Institute of Energy. (2011). *Strategic environmental assessment for the electricity plan between 2011-2015 and toward 2030*. Energy Institute: Ha Noi, Vietnam.

Institute of strategy and policy on resources and environment. (2009). *The status of resettlement programs in the hydropower dams in vietnam*. Ministry of natural resources and environment: Ha Noi.

Jayewardene, R., (2008). Can displacement be turned into development by compensation alone? . In M. M. Cernea & H. M. Mathur, (Ed.), *Can compensation prevent impoverishment? reforming resettlement through investments and benefit-sharing* (pp. 233-259). New Delhi: Oxford University Press.

Kim, A. M. (2004). A market without the ?right? property rights. *Economics of Transition, 12*(2), 275-305. doi:10.1111/j.0967-0750.2004.00179.x

Knetsch, J. L. (1983). *Property rights and compensation: Compulsory acquisition and other losses*. London: Butterworth.

Kusiluka, M. M., Kongela, S., Kusiluka, M. A., Karimuribo, E. D., & Kusiluka, L. J. M. (2011). The negative impact of land acquisition on indigenous communities' livelihood and environment in Tanzania. *Habitat International, 35*(1), 66-73. doi:http://dx.doi.org/10.1016/j.habitatint.2010.03.001

Larbi, W. O., Antwi, A., & Olomolaiye, P. (2004). Compulsory land acquisition in Ghana—policy and praxis. *Land use Policy, 21*(2), 115-127. doi:10.1016/j.landusepol.2003.09.004

Le, P. (2013). Corruption through "friendship groups". *The Law of Ho Chi Minh,*

Lerer, L. B., & Scudder, T. (1999). Health impacts of large dams. *Environmental Impact Assessment Review, 19*(2), 113-123. doi:http://dx.doi.org/10.1016/S0195-9255(98)00041-9

Marsh S.P., MacAulay T.G. and Hung P.V. (2007). In Sally P. Marsh, T. Gordon MacAulay and Pham Van Hung (Ed.), *Agriculture development and land policy in vietnam: Policy briefs.* Australia,: Australian Centre for International Agricultural Research.

Mathur, H. M. (2013). *Displacement and resettlement in India: The human cost of development* USA and Canada: Routledge.

McDonald-Wilmsen, B., & Webber, M. (2010). Dams and displacement: Raising the standards and broadening the research agenda. *Water Alternatives, 3*(2), 142-161.

Miceli, T.J. and Segerson, K. (2007). The economics of eminent domain: Private property, public use, and just compensation *Foundation and Trends in Microeconomics, 3*(4), 275-329.

MOIT, Ministry of Industry and Trade. (2013). *Result of hydropower dam construction survey nationwide.* Ministry of Industry and Trade: Ha Noi, Vietnam.

Niemann, P., & Shapiro, P. (2008). Efficiency and fairness: Compensation for takings. *International Review of Law and Economics, 28*(3), 157. Retrieved from http://search.proquest.com/docview/217423475?accountid=14772

Phan Vu Quynh Chi & Akimi Fujimoto. (2012). Land tenure and tenancy conditions in relation to rice production in three villages in the red river delta, vietnam *Issaas, 18*(No.1), 31-48.

Quang, N. N. (2002). Vietnam and the sustainable development of the mekong river basis. *Water Science and Technology : A Journal of the International Association on Water Pollution Research, 45*(11), 261-266.

Scodanibbio, L., & Mañez, G. (2005). The world commission on dams: A fundamental step towards integrated water resources management and poverty reduction? A pilot case in the lower Zambezi, Mozambique. *Physics and Chemistry of the Earth, Parts A/B/C, 30*(11–16), 976-983. doi:http://dx.doi.org/10.1016/j.pce.2005.08.045

Scudder, T., (1997). Social impacts of large dam projects In Dorcey, T., Steiner, A., Acreman, M., Orlando, B. (Ed.), *Large dams. learning from the past, looking at the future* (pp. 41-68). Washington, DC,: World Bank.

Scudder, T., (1997a). Social impacts. In Biswas,A. K. (Ed.), *Water resources: Environmental planning, management and development* (pp. 623-665). New York: McGraw Hill.

Scudder, T. (2001). The world commission on dams and the need for a new development paradigm. *International Journal of Water Resources Development, 17*(3), 329-341. Retrieved from http://www.ingentaconnect.com/content/routledg/cijw/2001/00000017/00000 003/art00005

Sebenius, J. K., Eiran, E., Feinberg, K. R., Cernea, M., & McGovern, F. (2005). Compensation schemes and dispute resolution mechanisms: Beyond the obvious. *Negotiation Journal, 21*(2), 231. Retrieved from http://search.proquest.com/docview/205191948?accountid=14772

Syagga, P. M., & Olima, W. H. A. (1996). The impact of compulsory land acquisition on displaced households: The case of the third Nairobi water supply project, Kenya. *Habitat International, 20*(1), 61-75. Retrieved from http://search.proquest.com/docview/38911610?accountid=14772

The Central hydropower join-stock company. (2011). *'The business and financial report'*. Website. Retrieved from http://www.chp.vn/Default.aspx?tabid=f91452b9-2cf8-424c-b798-9c60f0eeacd8

The government inspectorate of Vietnam. (2012). *The situation and result of citizen meetings, resolution of complaints and denunciations between 2008 and 2011and solutions for coming years.* 1198/BC-TTCP. The government inspectorate of Vietnam: Ha Noi, Vietnam.

The Thua Thien Hue provincial people's committee. (2005). Administration maps of the thua thien hue province. july, 2013, Retrieved from http://www1.thuathienhue.gov.vn/PortalNews/Views/Map.aspx

Thu, T. T., & Perera, R. (2011). Consequences of the two-price system for land in the land and housing market in ho chi minh city, vietnam. *Habitat International, 35*(1), 30-39. doi:http://dx.doi.org/10.1016/j.habitatint.2010.03.005

Thua Thien Hue People's Committee. (2013). Socio-economic development data of thua thien hue province in 2013. 05/22, 2014, Retrieved from http://www1.thuathienhue.gov.vn/portal_es/Views/Article.aspx?CMID=28&T LID=224

Turton, D. (2009). Can compensation prevent impoverishment? reforming resettlement through investments and benefit-sharing? Edited by Michael M. cernea and hari mohan mathur. *Development and Change, 40*(5), 983-984. doi:10.1111/j.1467-7660.2009.01595.x

Văn Thị Thanh, M. (2013). From "changing ways to work" to "some urgent issues of the party building at the present". *Communist Review,*

VIITANEN, K., & KAKULU, I. (2008). Global concerns in compulsory purchase and compensation prococess *FIG Working Week 2008 : Integrating Generations and FIG/UN-HABITAT Seminar : Improving Slum Conditions through Innovative Financing*Sweden 14-19 June 2008.

WCD. (2000). *Dams and development, a new framework for decision-making. the report of the world commission on dams* Earthscan Publications Ltd: London/Sterling, VA.

World Bank. (2004). *Involuntary resettlement sources book: Planning and implementation in development projects.* Washington, DC: The International Bank for Reconstruction and Development.

World Bank. (2011). *Compulsory land acquisition and voluntary land conversion in vietnam : The conceptual approach, land valuation and grievance redress mechanism.* Washington D.C: The Worldbank.

Yuefang, D., & Steil, S. (2003). China three gorges project resettlement: Policy, planning and implementation. *Journal of Refugee Studies, 16*(4), 422-443. Retrieved from http://search.proquest.com/docview/37844732?accountid=14772

Zaman, M. Q. (1996). Development and displacement in bangladesh: Toward a resettlement policy. *Asian Survey, 36*(7), 691-703. Retrieved from http://www.jstor.org/stable/2645717

Zoomers, A. (2010). Globalisation and the foreignisation of space: Seven processes driving the current global land grab. *The Journal of Peasant Studies, 37*(2), 429-447. doi:http://dx.doi.org/10.1080/03066151003595325

5 DAM-INDUCED DISPLACEMENT IN CENTRAL VIETNAM: VULNERABILITY, INEQUALITY, AND LIVELIHOOD PATHWAYS AFTER RESETTLEMENT TO THE VICINITY OF URBAN FRINGE [14]

5.1 Introduction

Dam-induced displacement and resettlement (DIDR) is neither new nor an unusual phenomenon, but it has becomes a reason for concern worldwide because of its speedily growing in scale and severe negative impacts, especially in developing countries (Cotula, et al., 2009; Mathur, 2013; Scudder, 2005). Large numbers of farmers have been displaced as a result in Africa, Latin America, Central Asia, Southeast Asia, and also in India and China (Cotula, et al., 2009). Hydropower dam construction greatly affects the livelihoods of forcibly displaced people and communities. Displacement and resettlement induced by hydropower dam development is one of the social pathologies that inevitably result from a modernization agenda aiming for renewable and sustainable energy development (Cernea, 2009). Although several studies acknowledge positive developments in displaced communities after being resettled (Agnes, et al., 2009; Nakayama, et al., 1999; Scudder, 2005), the majority report that DIDR causes great negative impacts for those affected people, even long-term (Gebre, 2003; Karimi and Taifur, 2013a; McDowell, 1996; Roy, 1999; Satyanarayan, 1999). DIDR leads to damage in production capacity, income, culture, and wellbeing (Bartolome, et al., 2000; M. Cernea, 2003; Scudder, 1997; Ty, et al., 2014; Wilmsen, 2011).One of many reasons leading to the downward development of displaced communities is the inappropriate selection of resettlement sites that do not enable displaced people to access better job opportunities after resettlement(WCD, 2000).

A review of available literature shows that DIDR programs typically focus on resettling displaced people to 'similar' settings, i.e., from rural-to-rural region and attempt to reconstruct similar livelihoods as those lost due to dislocation. This proximity principle in itself is logical as it strives to minimize changes in livelihood sources and setting and, as such, limits dislocation. Yet, at the same time, one may wonder how realistic such reconstruction of natural resource-intensive (land, water, forest) livelihoods is in a context of increasing scarcity of natural resources (land, water, forest). In many cases resettled people are left with less land and worse access to common pool resources than before resettlement or these resources are of a lesser quality (Ty, 2008). Recently, several countries have resettled displaced people from

---

[14] *This chapter will be submitted as* Ty, P. H., Zoomers, A., & Van Westen, A. C. M. Dam-induced displacement in central Vietnam: vulnerability, inequality, and livelihood pathways after resettlement to the vicinity of urban fringe. *Journal of Rural Studies*

rural-to-urban areas, as in the case of China (Wilmsen, 2011) and Indonesia (Yoshida et al., 2013). This model attempts to replace rural livelihoods by urban employment, but, here again, livelihood outcomes of displaced households are often poorer than before resettlement, possibly because people lack skills, networks and other assets necessary to do well in a very different environment (Wilmsen, 2011).

However, there are inequitable development patterns within communities because different families start from different positions of power and resource endowments, and therefore processes of upward and downward social mobility are less linear (Scoones&Wolmer, 2002). A substantial number of people once poor, became richer, and vice versa (Zoomers, 2010). It is consistent with the acknowledgement that livelihood is a dynamic process that requires longitudinal analysis to determine long-term adaptation or resilience of displaced households (Scoones, 2009). Meanwhile, research on impact assessment of dam-induced displacement and resettlement often measure consequences at two points, before and after displacement (Bui & Schreinemachers, 2011a; Wilmsen, 2011). Bui and Schreinemachers (2013) confirm that comparing two points only gives a snapshot of the changes and adaptation processes that often perceives negative impacts. Additionally, literature shows that resilience studies focus more on the impact of environmental and climate change than the impact of development activities. For example, (Adger, 2000) refers to resilience as the ability of groups or communities to cope with external stresses and disturbances as a result of social, political, and environmental change, but not the intervention of economic and energy development activities, such as hydropower dam development projects.

Therefore, the objective of this paper is to investigate the long-term adaptation strategies of displaced households due to hydropower dam construction in order to explore another option that has received little attention, i.e., resettlement from a rural area to a location in the interface between rural and urban spheres: a rural setting within easy access to urban employment and services. In this way, this paper aims to fill a gap in the literature on the role of location and livelihood reconstruction after resettlement. The case concerns a resettlement village displaced by hydropower dam construction in a mountainous region of Central Vietnam and relocated to the vicinity of Huế city. This study examines how this community reconstructed their livelihood, and why their livelihood pathways change as a result of resettlement, addressing the following sub-questions: firstly, to what extent is the displaced community vulnerable to impoverishment risks, and if present which livelihood measures they initiated to reduce vulnerability? Secondly, who are able or not able to adapt, and why? Thirdly, what are the implications for improved resettlement policies towards providing good livelihood opportunities for forcibly displaced people to sustain and obtain a better life? To answer these questions, we collected longitudinal data from three surveys in 2009, 2011, and 2014.

This paper is organized as follows. Firstly, the section following the introduction reviews the literature related to resettlement approaches to find pros and cons of those approaches and their implantation in practice. Secondly, from that, outcomes are explained to examine for alternatives. Thirdly, we overview hydropower dam development, displacement, and resettlement, and study findings on livelihood change of displaced communities after resettlement in Vietnam. Fourthly, the study methodology section provides information on the study site, conceptual framework, materials, and tools for data collection and analysis. Fifthly, the findings of the case study are described and analysed to answer research questions. Finally, we discuss the significance and implications of the results, and draw conclusions.

5.2 Livelihood pathways after dam-induced displacement

At the estimation of the International Commission on Large Dams (ICOLD), each year, over 10 million people are affected by development-induced displacement (Mcdonald, et al., 2008; World Bank, 2004), of which, four million are displaced by 300 large dams and six million uprooted by urbanization and transportation development projects (Cernea, 1995) In the 1990s, approximately 90 to 100 million people were forcibly displaced by land and water development projects (Cernea, 2000; WCD, 2000). It has been fourteen years since the report of the World Commission on Dams (2000), so we imply that roughly 100 million people have been displaced by similar projects. According to the WCD, (2000), displacement and resettlement could bring negative or positive development to affected people depending on resettlement approaches, practical enforcement of policies, and characteristics of displaced communities before and after being resettled.

According to the literature, displacement and resettlement is categorized into three approaches: managerial, movementist, and risks and rights approaches (Dwivedi, 2002). The first focuses on applied concerns and considers displacement as an inevitable and unintended outcome of development. This approach is appreciated by resettlement planners, managers, and applied academics that often search for solutions to minimize the adverse impacts of displacement by effective ways of designing and handling the appropriate legal, managerial, and policy framework (Dwivedi, 2002). Cernea, McDowell, and Picciotto are the most influential scholars who articulate this approach (Dwivedi, 2002). Cernea (2000) stressed that it is essential for not only studying disruptions but also helping formulate reconstructive strategies to turn displacement into development opportunities. Therefore, Cernea (1997) proposes the risks and reconstruction model for resettling displaced population serving. This approach is supported by the World Bank. However, this approach is top-down, making resettlement planners sensitive to local needs (Dwivedi, 2002) that often neglect the benefit of displaced people (Ty, et al., 2013).

A second approach offers another direction proposed by action research scholars who consider displacement as a manifestation of a crisis in development, and they work mainly on its causes. To them, displacement is evidence of development's uneven and unfair distribution of costs and benefits that do not improve people's well-being but destroy their existing ways of life (Dwivedi, 2002). As Parasuraman (1999), who lies closer to the movementist approach, criticizes that "development projects often favour the benefits of minority elite, while millions of people pay the price without reaping any benefits." However, this approach focuses much on community needs and initiatives while failing to link the displacement with broader national and global processes. Dwivedi (2002) says that Parasuraman appears indecisive, since his heart is dictated by the movementist and his mind by the managerial approach. Filling gaps between the two, WDC proposes the rights and risks approach, which combines the risks and reconstruction model and the rights articulated by the movementist approach. But representative of neither approach are happy with WCD's recommendations. Managerial scholars dismiss it as impractical and unimplementable, while movementists consider it as a necessary but not a sufficient step forward. For example, Medha Patka, a movementist, said that "even the rights recognised and risks assessed, and stakeholders identified, existing iniquitous power relations would too easily allow developers to dominate and distort such process" (WCD, 2000).

Thus, there is no perfect resettlement approach, and, therefore, the implementation of these approaches becomes more problematic in practice. The WCD study (2000) found that physical relocation often causes a loss of access to traditional means of livelihood, including agricultural production, fishing, livestock grazing, fuel wood gathering, and collection of forest products because it takes the form of forced and coercive displacement or, in a few cases, even killing. Further, compensation measures often fail to offset these damages (Cernea, et al., 2009; Fernandes, 2008; Jayewardene, 2008; Wilmsen, 2011). Bureaucratic compensation process often underestimates the value of lost land and other properties; delays in payment and lost livelihoods are rarely compensated (Mathur, 1999; Tan & Wang, 2003; Ty, et al., 2013; WCD, 2000). Displaced communities are often forced to resettle in regions with poor land and depleted natural resources. Resettlement plans often gives little consideration on the availability of livelihood opportunities or reference of displaced people themselves (WCD, 2000). Consequently, displaced people often become poorer, marginalized, and isolated (Bartolome, et al., 2000; Gebre, 2003; Karimi&Taifur, 2013b; McDowell, 1996; Roy, 1999; Satyanarayan, 1999; Scudder,

1997; WCD, 2000; Wilmsen, 2011), forcing them to leave resettlement sites and migrate (WCD, 2000).

In contrast to the disruption consequences of physical resettlement, some resettlements in Brazil, Ghana, and China benefit resettlers to make positive development. Particularly, China is considered as the first country to incorporate the notion of resettlement as a development opportunity with two innovative programs: the Partnership Support Programme and the Development Assistance Fund. In these models, the State Council redistributes income from power generation in the relocation areas to improve the living standards and livelihood opportunities for resettlers (Picciotto, et al., 2001; Wilmsen, 2011). In the case of the town of Zigui, Hubei province, China, resettlers benefitting from employment opportunities provided by local enterprises, saw their income significantly increased after relocation. But low education or lack of professional skills constrain them to be employed in these local enterprises (Wilmsen, 2011). Other resettlements improved the living standards of resettlers because they enable displaced people better access to land ownership titles, infrastructure, irrigation for farmland, health care, markets, banks, city centres, higher education, and job opportunities for the younger generation (Manatunge&Takesada, 2013; Sisinggih, et al., 2013; Souksavath& Nakayama, 2013). Therefore, WCD (2000) findings show that adequate laws, policies, plans, financing capacity, and political will of governments and project authorities could make resettlers benefit, but the capacity of resettlers may not correspond to the opportunities that are available.

Another important factor that influences the rehabilitation of displaced people are the characteristics of five vital livelihood capitals before and after resettlement, including natural, human, physical, financial, and social capital (Blaikie, et al., 1994; Chambers, 1995; Chambers& Conway,1992; De Haan, 2000). Households that have more land and natural resources could receive higher compensation and, therefore, start their resettlement in better circumstances than others (Karimi&Taifur, 2013c). As a consequence, resettlement might produce both winners and losers (Fujikura & Nakayama, 2013). However, the movement of household development is a dynamic phenomenon, so some could be winners in certain times but might be losers in other contexts (De Haan&Zoomers, 2005; Scoones, 2009). Therefore, we should study the livelihood pathways of displaced communities in the long-term to reduce the limitations of research findings that short-term studies cannot provide (Bui & Schreinemachers, 2011b; Wilmsen, 2011).

5.3 Hydropower dam development and resettlement policy in Vietnam

The survey of the Ministry of Industry and Trade of Vietnam (MOIT, 2013) showed that a significant increase in the number of hydropower dams. To date, 260 large-scale projects are operational, 211 plants are under construction to operate by 2017, and the rest are being licensed and registered. Additionally, 452 small-scale hydroelectricity plants are either operating or under construction across the country. Hydropower dam construction has displaced 44,557 households or approximately 200,000 people (Bui &Schreinemachers, 2011b) and expropriated 133,930 hectares of land (MOIT, 2013).A salient feature of this displacement is that some 90% of affected people in Vietnam belong to minority ethnic groups living in mountainous areas who rely on land and natural resources for their livelihood (Bui & Schreinemachers, 2011; Bui & Schreinemachers, 2013; Cao, 2003; CODE, 2010; Dao, 2010; Dao, 2011; Department of Co-operatives and Rural Development, 2007; Institute of Strategy and Policy on Resources and Environment, 2009).

Recently, many improvements have been made in the legal framework for displaced people, which have resulted in not only compensation for loss of houses, land, and properties on land but also in other assistance for agriculture production and job replacement. Decree 34/2010/QD-TTg of the government, issued in 2010, laid the first foundation for the improvement of compensation, displacement, assistance, and resettlement for irrigation and hydropower dam projects. The main objective is "to ensure displaced people to have the better or at least equal to living conditions before displacement," which is in line with the regulation of World Bank (2004). The first component of displacement and resettlement is compensation by lump-sum, cash payment for losses of land and other properties. In practice, it is often inadequate and delayed, and at an unfair, lowered value. The participation of displaced people is low and the process of compensation and resettlement is often not transparent. Project authorities often delay or do not keep promise to help displaced people after resettlement (CODE, 2010; Ty, et al., 2013). This is not a surprise because it takes the form of physical displacement in which land and cash compensation is the dominant principle (Wilmsen, 2011).

The second component is assistance for resettlement that is like a charity scheme, including supports for food, health care, education, fuel, electricity, agriculture production from one to two years. The principle of assistance aims mainly to recover agricultural production and forestry. Displaced people could receive technical training and finance to develop crop production and livestock models, but most resettlers prefer to receive in a lump-sum payment together with the compensation package and spend in non-productive ways (CODE, 2010; Ty, et al., 2014; VUSTA, 2006). Meanwhile, they often have access to limited or poor land and degraded natural resources for agriculture and forestry livelihoods. As a result, the majority cannot produce sufficient food and income for daily needs, and their living standards become worse than before displacement. For example, resettlers of the second largest Hoa Binh hydropower project in Vietnam remain poor 30 years after resettlement (CODE, 2010).

5.4   Research methods

This study applied the Impoverishment Risks and Reconstruction (IRR) model of Cernea (1997) and McDowell (2002) to evaluate vulnerability after resettlement. Additionally, the sustainable livelihood framework was used for designing the research and exploring livelihood adaptation strategies, outcomes, and influential factors on livelihood pathways and outcomes (DFID, 2001; Ellis, 2000; Scoones, 1998b). Inequitable distribution of income and land were identified for six groups: women-head, handicapped, young, Kinh, Co tu, and leader households by using Gini and Robin Hood indices[15]. Both quantitative and qualitative methods were applied to determine livelihood outcomes and to explain correctly reasons causing changes of livelihood pathways before and after resettlement. To increase the credibility and validity of the results, we applied triangulation approaches for collecting data. We focused on data and methodological triangulation because our experiences when interviewing displaced communities in Central Vietnam show that displaced people often give negative attitudes on displacement and resettlement. They often exaggerate living conditions before displacement while undervaluing outcomes after resettlement. Additionally, different stakeholders related to displacement and resettlement programs normally provide various information, sometimes opposite views and data.

In this study, we collected data in 2009, 2011, and 2014. Qualitative surveys started with key informant interviews with representatives of commune and district government, village leaders, and the patriarch to investigate initial views about the displacement and resettlement of Bo Hon village in different years. A focus group of 12 people was gathered, including four village leaders, the patriarch, and seven representatives of all households, to discuss the history of the village and major changes of livelihoods before and after resettlement. Normally, we had more people in focus group discussions than invited, who came and enthusiastically answered our questions because their houses are very close to each other. Group discussions with children were conducted understand their views about displacement, migration of children, culture change of younger generation, and economic conditions in their families. It revealed that children gave very useful information. We also had

---

[15] See (Maio, 2007) to discover definitions and formulas of Gini and Robin hood indices

observations and informal talks with villagers to discover their personal views about resettlement and livelihood changes. After that, 40 households, approximately 75% of the total population, were randomly selected from the list provided by the vice-village leader to evaluate the resettlement program and to quantify changes of household capital and livelihood outcomes to describe the pathway of livelihoods after 8 years resettlement. Following that, we undertook in-depth interviews with several households to discover insights of livelihood strategies. Paired sample T-test was used to compare mean income differences, and Pearson correlation was used to determine the relationship between income and other influential variables on livelihood changes.

5.5   The context of research site

This study investigated the livelihood change of a resettlement village displaced by Binh Dien hydropower dam in Thua Thien Hue province, Central Vietnam. The province has a population of over 1.1 million living equally in rural and urban areas. The economy depends mostly on the service sector (54%), industry (35%), while agriculture (10.3%). Hue city, a former capital of Vietnam, is the centre of the province with a population of 0.33 million, with many international heritage sites that make it the tourism destination of Vietnam.

The province has three important watersheds: Huong, O Lau, and A Sap river watershed. Huong river watershed is the largest with 2.830 km² covering three fourths of the province's area. The water resource potential led to hydropower development. The province constructed five dams and plans to build an additional six dams in coming years. Binh Dien hydropower dam was the first constructed in 2005 on the right tributary of Huong river watershed in Huong Tra district. The cost was 1.1 trillion VND[16] ($5.3 million USD) invested by Binh Dien Hydropower Joint stock, Ltd. It is a multipurpose hydropower plant, producing electricity, controlling flood and drought, and providing irrigation for agriculture. It started operation in 2009 and produced an electricity production of 181 million kWh to the National Grid power in 2011 (Binh Dien Hydropower Joint Stock Company, 2011).

For the construction of the Binh Dien hydropower dam, 616 hectares of land was acquired for its reservoir, which included 140 hectares of expropriated land, encompassing the whole village of Bồ Hòn. In 2003, Bồ Hòn villagers were told to resettle in a new village before the actual construction date of Binh Dien hydropower dam. In August 2006, the entire village received compensation and resettled in the new Bo Hon resettlement village, the Binh Thanh commune (Binh Thanh commune, 2008).

---

[16] 16,055 VND = 1 USD according to the 359 /TB – BTC of Ministry of Finance in 2006

History of the old Bồ Hòn hamlet (bản). The former Bo Hon hamlet originated from a group of 33 Cơ tu[17] households of the Huong Nguyen Commune before 1984. Hamlet (bản) is often used for a minority ethnic group living in mountainous regions (citation). Between 1984 and 1989, they moved to Lác River. Because of the 1995 flood, they moved to Bồ Hòn hamlet named after popular Bồ Hòn trees along the Huu Trach River. At that time, five Kinh[18] households moved in and lived together with them due to the flood, as well. The village was located in a valley of the Nam Hoa State Forest Enterprise, approximately 15 km from the centre of Binh Thanh commune and 40km from Hue city (See Figure 5-1). In the old village, Cơ tu accessed river banks to plant Lồ ô bamboo and practiced *slash-and-burn cultivation*. The main mechanism of access to land and other natural resources was under the customary law set by the patriarch and all villagers. Kinh had no right to access communal lands, but could access natural forests to exploit non-timber products. The infrastructure was very poor; the only way to go out of the village was by waterways. Also they had no electricity and water supply, and, without a school, most were illiterate (Focus group discussion, 2009).

The new Bo Hon resettlement village (thôn). The whole village was displaced and resettled in 2006. The hamlet was change to a village. The hydropower company constructed the resettlement village downstream of the dam and about 15km from the old village. It is closer to both the centre of Binh Thanh commune by a 2-km concrete road and to Hue city, 25km away by a well-connected road network. In the village, there is a communal house, a primary school, and a kindergarten. Access to a secondary school and high school is 4 km away in Binh Dien commune (See Figure 5-1). All households have electricity and water supply. Each household has a piece of land with a house, garden, and crop production in front of the house. Four neighbouring villages are nearby, and most of them are Kinh people. In 2013, there were 54 households with 278 people in the village; over 90% were Cotu people; and the rest is Kinh. Labours whose ages were under 45 years old comprise nearly 60%. Most people completed education from primary to high school, while only 12.5%were illiterate. Two people completed higher education after resettlement (Household surveys, 2014).

---

[17] In Vietnam, Cơ tu is a minority ethnic group, there are 61 thousand people, the 25th largest population among 54 ethnic groups in Vietnam, in which 90% live in the countryside and 5% reside in urban areas

[18] Kinh is the majority people in Vietnam accounting for 86% of population

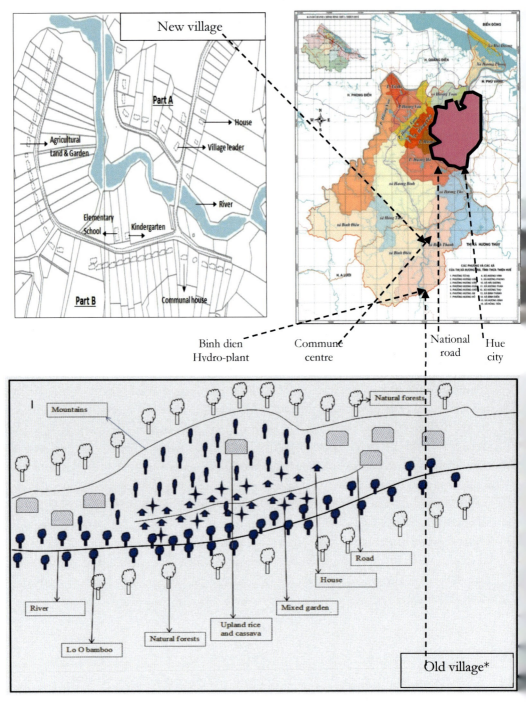

Figure 5-1 Location of Bo Hon village before and after displacement.
Source: Binh Thanh commune (2008); Centre for resource and environmental survey (2013);
Authors' participatory mapping, 2009 (redrawing)

106

5.6 Results

5.6.1 Compensation and resettlement plan: satisfactory to what extent?

According to decision 3721/2005/QĐ-UBND of Thua Thien Hue province, the compensation principle stipulated land for land and cash compensation for losses. Each household received a piece of land (0.3ha per household) attached to a house, and the majority (92.5%) of households received cash compensation for their losses. On average, each household received VND 35.8 million, in which 10%receivedmore than VND 100 million per household. Besides payment in cash, displaced households received other assistance for one year, but were paid in cash together with the compensation package (see more in Table 5-1). However, land was not compensated. According to the Land Law 2003, the project authority only recompensed legal lands with land use right certificates (Red Book[19]), whereas the displaced village practiced slash and burn cultivation and used land under the customary right system[20] without the official recognition of responsible local governments. For example, 61 hectares of Lồ ô bamboo land along river banks were not compensated because these trees were planted according to customary rule of the village without Red Books. Only Lồ ô trees over land were estimated to have been compensated for. Thus, customary rights were not taken into account. Additionally, the majority of displaced households were disappointed about the compensation process because it was unfair, and some households had more compensation than others. Particularly, nepotism and corruption were also found in this case (Michelle, 2011). After compensation and assistance, hydropower did not provide any further benefit sharing from power generation. Displaced household struggled to survive with a few supports from local authorities and NGOs. Only Lồ ô trees over land were estimated to compensate for. Thus, customary rights were not taken into account. Additionally, the majority of displaced households were disappointed about the compensation process because it was unfair, and some households had much compensation than others. Particularly, nepotism and corruption were also found in this case (Michelle, 2011). After compensation and assistance, hydropower did not provide any further benefit sharing from power generation. Displaced household struggled to survive with a few supports from local authorities and NGOs.

---

[19] Red book refers to the land use right certificate granted by competent government agencies that authorise to rectify the land use title of individuals, households, and organizations. The cover of land use right certificate is red, which is why people normally call it the red book.

[20] In the former village, the expansion of land depended very much on labour force and the boundary was determined by local perception among villagers with the approval of patriarch and village headmen.

Table 5-1 Compensation and assistance for displacement – whole village in 2006. *Source: Binh Thanh commune, 2008; * 16,055 VND = 1 USD in 2006*

Categories of compensation and assistance	Unit	Quantity	Compensation rate per unit (million VND*)	Total compensation (million VND)
Compensation for loss of trees and crops				
- Forest trees	ha	10	2.1	21.0
- Dry crops	ha	7.84	5.8	45.5
- Lồ Ô bamboo trees	ha	60	7.5	450.0
Total in cash				*516.5*
Compensation per household				*10. 8*
Assistance for resettlement				
-Reclaiming land for industrial crops	ha	6.66	3.0	19.9
-Reclaiming land for rice cultivation	ha	3.2	8.0	25.6
- Agricultural development	household	48	5.0	40.0
- Garden improvement	ha	2.67	4.0	10.7
-Forest and agricultural extension	household	6	4.0	24.0
- Job trainings	household	48	1.5	72.0
- Livestock	household	48	1.5	72.0
- Forest plantation	ha	25.35	9.0	228.2
- Red book fees	household	48	0.5	24.0
-Rice (21kg/person/12 months)	persons	193	3.8	184.8
- Health care and education	household	48	100	4.8
-Transportation of former houses	household	48	4.0	192.0
-Re-construction of cemetery	cemetery	65	0.5	32.0
- Support for the poor	household	48	1.5	72.0
Total assistance in cash				*1,202,527*
Support per household				*25.5*

108

5.6.2 Variation of risks and vulnerabilities

Cernea (1997) identifies eight means to an increased displacement probability of impoverishment: landlessness, homelessness, marginalization, increased morbidity and mortality, food insecurity, loss of access to common property, and social disarticulation. Downing (1996) also identified additional risks, including loss of access to public services, disruption of formal education activities, and loss of civil and human rights. However, according to Stanley (2004), displaced people are impoverished differently. Vulnerabilities occur when affected people are unable to counterbalance adverse impacts to cope short-term and adapt long-term (Adger, 2006; Cutter, et al., 2008; Luers, et al., 2003; Scoones, 1998b). Social vulnerability refers to disruption of livelihood and loss of security of individuals or groups who perceived risks. For vulnerable groups, the degree of vulnerability is pervasive (Chambers, 1995). In our research, we identified vulnerabilities after 8 years resettlement and found that loss of productive land for upland rice and dry crops was the greatest challenge for displaced people because they could not find replacement agricultural lands. According to resettlers, they had access to large track of fertile lands for crop production on hills and along the riverbanks in the old village, but very limited access to productive land for crop production after resettlement. Each household lost 87% of productive land, from 1.6 to 0.18 hectares, after resettlement (See Figure 5-2). Soil quality was also poorer than before resettlement. Over 90% of households said that traditional cassava became bitter and indigestible because of shallow soil depth and poor fertility.

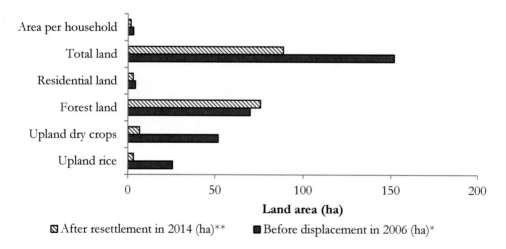

Figure 5-2 Land of Bo Hon village before and after resettlement. *Source: Binh Thanh CPC, 2006; Centre for resource and environmental survey, 2013; * n = 48 households; ** n = 56 households*

109

Amongst households, women-headed and handicapped households were more susceptible to land loss. After resettlement, women-headed households lost 85 per cent of land. Handicapped households lost 75 per cent of land and owned less threefold than other non-disabled households. Kinh households lost 84 per cent and owned less Cơ tu than six times. Households of village leaders owned double land than others both before and after resettlement. The differentiation in land distribution influenced significantly on compensation and livelihood restoration. Former and new village leaders had much higher compensation and then invested in reclaiming lands for Acacia forest plantation. They accumulated more land thanks to high financial capacity and therefore resulted in high inequality of land distribution in the village (see more in Table 5-2). This could be also explained by the substantial increase of Gini and Robin Hood indices of land distribution, they doubled after resettlement, from 0.33 and 0.24 to 0.6 and 0.48 respectively. This was really high inequality. Additionally, displaced households were very vulnerable to access to common pool resources because they lost access to 450 hectares of natural forests and water bodies after resettlement. They lost 90 per cent of fishing products and 97 per cent of rattan. As a result, each household lost nearly 55 per cent income, from VND 1.9 to 0.86 million a year. Kinh households were more susceptible than Co tu because they lost 80 per cent of their income.

As a consequence of land loss and loss of access to common pool resources, displaced people also lost their traditional livelihoods, in which the loss of Lồ ô bamboo plantation activity was the most challenge because each household owned 1.3 hectares and several households owned over 10ha before displacement. They could earn VND 15 million ($US700) per household a year from such production. Lồ ô bamboo was a sustainable income because it could re-grow every year, and its price increased significantly. In 2004, the price was VND 4,000 per trunk and doubled in 2009, then increased again in 2013 to VND 24,000 per trunk. They also could easily sell to retailers who went directly to the village by electric boats. After the land was flooded for the reservoir, they could no longer have income of Lồ ô bamboo.

Food insecurity was the most severe and long-term vulnerability to displaced people in the new village because they did not have productive land for cultivating rice and cassava crops. The annual productivity of rice declined significantly from 52kg to 18kg per household. Over 56% of households had to spend compensation and income to buy rice and food in the new village; whereas 70% could supply themselves in the former village.

Table 5-2 Land distribution between social groups before and after displacement (n=40 households)

Source: Household survey 2009 and 2014

Household groups		Land before displacement 2006 (ha)						Land in 2014 (ha)					
		mean	min	Max	range	SD*	sum	mean	min	max	range	SD*	sum
Household head	Female	2.4	0.5	4.2	3.7	1.4	14.5	0.4	0.3	1.2	0.8	0.3	2.6
	Male	3.7	0.8	13.6	12.8	2.5	124.4	2.1	0.3	9.5	9.2	2.5	73.7
Age group	<45	3.1	0.5	6.5	6.0	1.6	51.9	1.4	0.3	7.2	6.9	1.9	33.6
	>45	3.9	0.8	13.6	12.8	2.8	87.0	2.5	0.3	9.5	9.2	2.9	42.7
Ethnicity	Kinh	1.5	0.5	2.0	1.5	0.7	6.0	0.3	0.3	0.3	0.0	0.0	0.6
	Cotu	3.7	0.8	13.6	12.8	2.4	132.9	1.9	0.3	9.5	9.2	2.5	75.7
Health status	Normal	3.5	0.5	13.6	13.1	2.4	128.4	2.0	0.3	9.5	9.2	2.5	74.3
	Disabled	3.5	1.5	6.5	5.0	2.6	10.5	0.7	0.3	0.9	0.6	0.3	2.0
Social status	Not	3.1	0.5	6.8	6.3	1.7	107.1	1.6	0.3	9.5	9.2	2.2	54.4
	Leader	6.4	3.4	13.6	10.2	4.2	31.8	3.7	0.5	7.0	6.6	3.0	22.0
Total land		3.5	0.5	13.6	13.1	2.4	138.9	1.9	0.3	9.5	9.2	2.4	76.4

Notes: (*) is the standard deviation

As a result of this significant reduction in economic and food conditions, they could not maintain their culture and religion. As confirmed by 40% of households, they almost changed to the culture of Kinh people after resettlement: they spoke Kinh (Vietnamese) language frequently, built houses, prepared foods, wore clothes, gave names for children as Kinh people did. Several households changed from a Cơ tu religious tradition to Buddhism and Christianity because of the hunger relief received from delegates of these religions. Nearly half worshiped their ancestors and organized wedding ceremonies in the ways of Kinh people. Nevertheless, 60% still continued strong community interaction. For example, Cơ tu girls retained marrying with Cơ tu men. One girl, sixteen years old studying at 10th grade, said that her sister married a Cơ tu man in Nam Dong district because he had land for a rubber tree plantation.

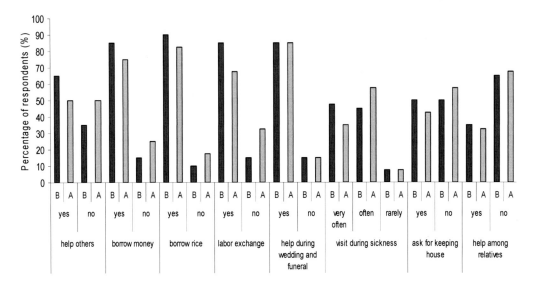

Figure 5-3 Community coherence before and after displacement (n=40 households). *Source: Authors' household survey, 2014. Note: B – before displacement; A – after displacement Source:*

In addition to the change of culture, the coherence of Bo Hon village gradually decreased. We found that 83% of households remained well the community interaction, but 17% did not see cooperation and mutual help anymore (See more in Figure 5-3). The patriarch said that many households lived more separately or independently since they did not trust each other as previously. Therefore, they did not ask neighbours to keep their houses as in the old village. The disintegration of the village happened because few collective activities were organized in the new village. The transformation of village management system from a customary to a government-oriented regime also increased incoherence. The role of the patriarch was no longer

112

important, in spite of the fact that a new village management board appointed by the commune authority played a more important role; it could not prevent conflicts that emerged in the new setting. Last but not least, unfair compensation and larger income gap reduced trust among households.

5.6.3   Responses to reduce vulnerabilities

The process of responding to reduce vulnerabilities could be expressed as a resilience process or adaptation. There are many similarities between resilience and adaptation, but they can be differentiated. Resilience refers to the degree to which a complex system is capable of self-organization (versus lack of organization or organization forced by external factors) that it is able to build capacity for learning (Carpenter, et al., 2001; Folke, et al., 2002) and recovering from adverse conditions, countering the effects of risks or shocks, and reorganizing a society to meet new conditions (Berkes, et al., 2003; Cutter, et al., 2008; Folke, et al., 2002). Further, adaptation can be considered as a continuous resilience in the long-term to secure well-being and reduce other new vulnerabilities (Davies &Hossain, 1997). Adaptation may be positive or negative: positive if it is by choice, reversible, and increases security and negative if it is of necessity, irreversible, and fails to reduce vulnerability. Different types of shock or stress, in turn, may result in different responses, including avoidance, repartitioning, resistance, or tolerance mechanisms (Payne &Lipton, 1994). Resilience and adaptation responses also might function as a mechanism of vulnerability transfer and/or transformation (Sapountzaki, 2007). In this case study, we found seven main strategies applied by resettlers as follows.

Spending compensation money: Most displaced households spent compensation mitigating food insecurity and income loss during initial years after resettlement. They used 20% of compensation for foods, 21% for savings but withdrew all for health care, school fees, food, and water and electricity charges in 2009. Half of compensation was used in buying furniture, motorbikes, mobile phones, TVs, refrigerators, and extending houses after construction by the hydropower company. A small percentage of the compensation was used for livestock and land reclamation for Acacia forest plantation. The majority of households confirmed that compensation enabled them to overcome the severe shortage of food and improve their living standards after displacement. In which, motorbikes and mobile phones allow them to communicate better with outsiders to get job opportunities in the region and with their families when they migrate to work in cities.

Restoration of traditional livelihoods: Along with compensation spending, displaced households began recovering traditional livelihoods in the new village right after resettlement. They replanted Lồ ô bamboo, but could not find suitable lands. Instead,

113

in 2007 and 2008 five households returned to the old village two times per month by electric boat to harvest bamboo trees not totally flooded. From this, each household earned VND 7 million per year. In 2009, they stopped harvesting Lồ ô because the hydropower company restricted access to the reservoir. In 2010, the Centre for Rural Development (CRD) of Hue University of Agriculture and Forestry (HUAF) came and supported the villagers to being replanting Lồ ô again. They facilitated the Huong River Protection Forest Management Board (HRPFMB) to share their 40 hectares land for Lồ ô bamboo plantation (Ty, et al., 2010). However, the village leader said that land was not officially relocated to them and quite far from their village, so that they did not take care of Lồ ô bamboo well. As a result, HRPFMB retook the land, and Lồ ô bamboo livelihood practically disappeared.

They also failed to replant traditional cassava, which was an important and favourable food. After resettlement, all households began growing traditional cassava in their garden, but poor soil quality prevented cassava from growing well. It became bitter and indigestible. In 2009, they stopped growing traditional cassava and changed to a new variety called KM94 industrial cassava. This idea came from a former leader who planted KM94 cassava at first. He said, "Industrial cassava was not good to eat but it produced higher productivity and its price was good, easy to sell. After that, I encouraged other households to plant KM94 instead" (Authors' survey, 2009).Ten households planted KM94 and each household earned 700 thousand VND in 2011 and 1.4 million VND in 2013. Nevertheless, the limited garden land restricted them to produce more cassava, while the price of cassava remained quite stable year-over-year (Authors' survey, 2014).

Resettlers also could not restore income from common pool resources. After resettlement, they looked for alternative natural forests and rivers in the region to collect non-timber products, such as rattan, honey, and Lantenier leaves for making hats, as well as for fishing. However, only six households continued this activity and each household generated only 0.9 million VND per month, which included two women-headed households.

Crop diversification: Because of the poor and limited land, displaced household were forced to cultivate other crops, including lemon grass, pineapple, and taro. They explained that these cash crops could grow on poor soils and required less land than other crops. In addition, it was very easy to sell to retailers who visited the village every day for business. In 2013, twenty-one households planted cash crops, in which women participated in this activity the most. Although income from cash crops was not high, VND 1.7 million per household per year, and they said that it was very important for buying daily foods. They also exchanged with retailers lemon grass, taro trunk, and pineapple for other foods. Over 90% said that they would continue growing lemon, taro, and pineapple in coming years (Authors' survey, 2014).

Reclaim lands for Acacia forest plantation: After resettlement, the authority allowed resettlers to reclaim unused uplands[21] to plant Acacia forests that were considered as the most important economic forest in the region[22]. In 2007, twenty-seven households started planting Acacia forests. Subsequently, JBIC and WB3 projects came and supported technical trainings and loans with low interest to invest in Acacia forest plantation. They also assisted local authorities to allocate forest Red Books to resettlers. In 2013, all households received Red Books for 75.8 hectares of Acacia forest (Centre for Resource and Environmental Survey, 2013). On average, each household had 1.65 ha of Acacia forest. Most started selling Acacia forest products[23], and each household earned about VND 8 million per year (See Table 5-3). The income from Acacia forest became the second largest source after labour wage income (See Figure 5-4). Yearly Acacia planting allows for successive yearly income. Most households confirmed that they will continue planting Acacia forest if they have more land because Acacia trees yield high income over other trees. This could be explained more in the findings of Ha (2013).[24]

---

[21] Unused uplands are lands not allocated to any land users and under the management of commune authority

[22] Acacia forest is a very important product of Vietnam that exported 2 mil m^3 per year of Acacia wood pulp materials to Japan, China, Taiwan, and Korea. It is the fourth largest export country worldwide (Wood Resources International LLC, April 16, 2009). In Thua Thien Hue province, Acacia forests have significantly increased since 2000, from 307 hectares to 23,839 hectares in 2010 which comprised 90% of the production forests of the province. In which, over 65% of Acacia forest are owned by individuals and households. According to the forest protection and development plan from 2011 to 2015, there is an increase of 2,852 hectares of Acacia forest annually

[23] Resettlers sell Acacia forest products to retailers living in the same commune or in Hue city. They come directly to the village to negotiate the price for Acacia forest producers, if agreed, retailers hired people in the village to harvest and then transport to the wood pulp production companies in Chan May port, Thua Thien Hue province.

[24] Ha, (2013) investigated factors influencing the famers' decision on Acacia plantation and confirmed that the development of Acacia forest was based on three main reasons. First, the local government and many international organizations provided many technical and financial supports for farmers to plant Acacia in the long-term, including ADB, SNV, ETSP, WB, and JBIC. In addition, these organizations facilitated local authorities to allocate land and grant land use rights certificates so that farmers were secured to invest on their lands. Second, Acacia forest often created a high profit for farmers because it grows faster than other forest species. After 5 to 7 years, each hectare of Acacia forest could yield from 1000 to 2000 USD after subtracting 250 to 500 USD costs for planting and harvesting. More importantly, the price and market was very stable with an annual increase of 12%. Finally, Acacia forest was less damaged by natural disasters, such as flood, droughts, and hurricanes, than other trees. Even if broken, it also could be sold to the wood pulp processing factories.

Table 5-3 Acacia forest land and income in 2014 (n = 40 households). *Source: Authors' household survey, 2014; Binh Thanh CPC, 2013*

	Min	Max	Sum	Mean	Range	SD
Acacia forest land (ha)	0.0	9.3	62.6	1.65	9.3	2.4
Annual income of Acacia forest (million VND per household)	0.0	45.0	317.3	7.90	45.0	12.4

Table 5-4 Acacia forest land and income among household groups (n=40 households). *Source: Authors' household survey, 2014*

Household groups		Acacia forest land (ha per household)	Annual income of Acacia forest (million VND per household)
Household head	Women-headed	0.02	0.17
	Men	1.81	9.21
Ethnicity	Cơ tu	1.71	8.73
	kinh	0.00	0.00
Health capability	Disabled	0.67	2.80
	not	1.61	8.27
Age group	<45	1.19	6.25
	>45	2.01	10.03
Social status	Leader	3.78	18.25
	not	1.14	6,.2
All households		*1.65*	*7.90*

However, several vulnerable households, including women-headed, handicapped, and Kinh households, had less land and income than others. They explained that plantation of Acacia required quite high costs and labour, and they did not have enough money for this planting. Further, their children migrated to work in other cities, so they could not reclaim land for Acacia forest. Households of village leaders accumulated much more land for Acacia forests than other in the new village (See Table 5-4).

Working for Acacia forest owners in the region: As mentioned, Bo Hon village was located in the densely planted Acacia forests and close to the points of purchase and transportation of Acacia wood pulp to processing factories. As a result, this was a good opportunity for displaced people to work for other Acacia forest owners. Since 2008, 24 households with 35 labours have been involved in this activity. Both men

116

and women between 14 and 45 years worked for Acacia forest owners. They often went to work by motorbike and contacted forest owners via mobile phones they purchased after displacement. Half of them confirmed that the demand was high, but the rest said that this job was not stable especially in the rainy season. However, it was the largest income source in 2013 when each household earned 19 million VND per year (See Figure 8). Women-headed, disabled, and Kinh households benefited most from this activity since their children earned a significant income, improving their living standards. Nonetheless, most of households confirmed that Acacia forest plantation and harvesting was very hard work. They will continue working, but will send their children to Ho Chi Minh City for working in textile and garment factories (Household survey, 2014).

Worker migration to cities and industrial zones: Migration was another non-farm livelihood initiated in 2011 by10 migrants and 13 migrants in 2013. Most worked for textile and garment factories in Ho Chi Minh City, the products of which are the most important export of Vietnam (Tien, 2013)[25]. Only two migrants worked in the industrial zone in Hue city. They were from 15 to 23 years and had completed primary to high school. Over 80% are men. In 2011, each household of migrants received over 10 million VND of remittance, but about 4 million VND in 2013. The reason for his decline was the economic crisis in 2012 influenced the growth of textile and garment industry and reflected directly on their salary (Author's in-depth interview, 2014). The majority confirmed that they spent 70% of remittance for daily food and 30% for health care, education, and buying household facilities. Thus, remittance-assisted displaced people reduced the vulnerability of food insecurity after resettlement. However, there was no program supporting them to find migration work; they found work themselves at first and then introduced this employment to others in the village. They established several groups and stayed together for the whole year, only returning to the village on the occasion of the New Year holiday. The village leader who has three migrants in Ho Chi Minh city said that salary was quite low, from 2.5 to 3.5 million VND per month, so it was only enough to feed themselves and, therefore, did not have money to go back often. More importantly, the group discussion with children showed that more children dropped secondary school to migrate. This might cause another risk and vulnerability to disruption of formal education as identified by Downing (1996).

Accept new culture and religion: We found that 40% of Cotu households accepted other cultures and new life styles easily, particularly from Kinh people. One Cotu man,

---

[25] The value of textile and garment export sector reached over 20 billion USD in 2013 with an annual growth of 15% and employed 2 million workers. Every one billion increase of textile export creates 100 hundred thousand jobs. Most workers migrate from rural areas.

117

35, explained that he had many Kinh friends in the commune since he worked with them for Acacia forest owners and communicated with them, and, therefore, he spoke Kinh language frequently. Another girl in the children's group discussion (16 years old studying in 10th grade) said that she studied with Kinh classmates. At first, she felt nervous because their friends laughed at her as a minority. However, they eventually stopped, and she felt that she was no longer significantly different from her classmates. We also asked the children group "Did you parents take you to Hue city often?": 50% said that their parents sometimes took them to Hue city to buy clothes, while others did not go to Hue city at all because their parents did not have money. We found that over 80% wanted to be Kinh in future, so we asked the children's group "Why did you want to be a Kinh?" They said that Kinh people are smarter and richer, so it is better to become like them to have better life. During discussion with the children, we also witnessed several Kinh songs. The former leader, 70, predicted that this change would be clearer in the near future, as many Cơ tu continue migrating to Ho Chi Minh City for work and to Hue City to studying.

5.6.4   Livelihood outcomes, inequality, and pathways

Displaced households applied many strategies to reduce vulnerabilities. The question is whether these responses enhance livelihood outcomes that are often characterised by improved incomes, reduced poverty and vulnerability, increased capabilities and better food security, sustainability of resource base, and enhanced stocks of livelihood assets for future livelihood strategies (DFID, 1999; Scoones, 1998a). In this study, we examined changes in income, poverty, food security, and vulnerability. We found that displaced households experienced two stages after 8 years of relocation: firstly, disruption between 2006 and 2009 and, secondly, recovery from 2009-2013 (See Figure 5-4). In the first stage, income decreased tremendously from 665 to 130 million VND. Paired sample T-test also confirmed that annual income per household dropped significantly from 2006 (m= VND 15.6, SD=17.8 million a household per year) to 2009 (m= VND 3.2, SD=2.4, t (39) =−4.34, p=0.00009 two tailed), in which income from Lồ ô bamboo forest and common pool resources almost disappeared. Instead, they started working for Acacia forest owners and selling their cash crops. They also began reclaiming lands and planting Acacia forests.

In the second stage, income of displaced households increased substantially from 130 to903millionVND, where wage labour income and Acacia forest plantations played the most important part in total. Remittance of migrants also began increasing and surpassed the income of agriculture and common pool resources (See Figure 5-4). Comparing income before and after displacement by using paired sample T-test revealed that income in 2013 (m=23.8, SD = 39.7 million VND a household per year)

118

was not statically higher than before displacement in 2006 (m= 15.6, SD=17.8 million VND a household per year, t (39) =1.25, p=0.220 two tailed). We supposed that it was a fast recovery after resettlement because many communities displaced hydropower dams, for example displaced people due to the construction of Hoa Binh dam, became poor or poorer for over 15 years after resettlement (Cao, 2003; CODE, 2010; Dao, 2010).

The fast recovery of income was closely associated with Acacia forest plantation land and wage labours. As Pearson correction analysis revealed that income recovery was significantly correlated with income of wage labour (r=0.577, p= 0.00009 two tailed, N=40) with Acacia forest land size after resettlement (r=0.645, p= 0, 000007 two tailed, N=40), and with income of Acacia forests (r=0.586, p= 0.00015 two tailed, N=40). This suggested that households had more income if they had more land for Acacia forest plantation and wage labours after resettlement. Although the number of migrants significantly increased, the contribution of remittance was lower than other factors (See Figure 5-4).

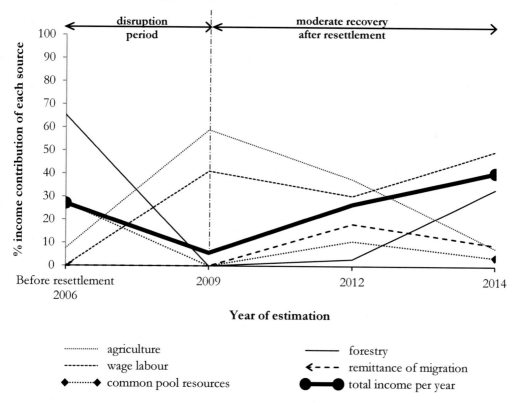

Figure 5-4 Income changes of displaced households before and after displacement (n=40) *Source: Authors' household survey 2009, 2011, and 2014*

119

But did the income recovery help reduce poverty? In order to estimate poverty rates, we converted nominal incomes to real incomes for each year using CPI indices in rural areas of Thua Thien Hue province. After that, real income of each household was compared with the poverty threshold of Thua Thien Hue province corresponding to certain years[26] to estimate the percentage of poor households. The result revealed that poverty rate increased sharply from 32.5% in 2006 to 62.5% in 2009, but reduced by half in 2013. There was a slight reduction of poverty rate before and after resettlement. However, it was still much higher than the poverty rate of Thua Thien Hue in 2013, which was 12.7% in rural areas and 14%in minority ethnic communities. Particularly, it was much greater than the poverty rate of whole province in 2013, which was 6.5% (Thua Thien Hue People's Committee, 2013). Thus, Bo Hon displaced households made a great recovery, but it could not match the rapid growth of the whole province.

However, there was a considerable variation among households in the process of recovery after resettlement. We found that over 55% of households lost income, but 45% gained more income in 2011; over 42% lost while 58% gained income in 2013. Before resettlement in 2006, six households that did not have income at all, including two women-headed households, one Kinh, two disabled and one old household. However, women-head and handicapped households made a good improvement in 2013, but elderly households could not because they had no land for Acacia plantation, income from wage labour, and remittance of migration (See more in Table 5-5). Young households made a moderate increase in income because they earned from working for Acacia forest owners, gardening, migration, and their own Acacia forests. The strong recovery of income enabled 50% of households to become better off and 7 households to escape from poverty completely. However, 6 households stayed poor and 5 families became poorer than before displacement. It can be seen from this analysis that livelihood outcomes were different among households, and its movement was not linear. Many lost income and moved downward while others gained and moved upward. It was consistent with the findings of Zoomers and Albó (2000).

---

[26] According to the poverty standard of Vietnam between 2006 and 2010, the poverty line was between 0.26 and 0.4 mil VND per month each household. However, according to the poverty standard of Vietnam between 2011 and 2015, the poverty line increased to VND 0.41 and 0.5 mil per month each household between 2011 and 2015.

Table 5-5 Variation of income among household groups before and after resettlement (n=40 households).

Source: Authors' household survey, 2009 and 2014. Unit: million VND

Household groups		Mean		Min		Max		Standard deviation		Sum	
		before	after	before	after	before	after	before	after	before	after
Household head	Women	11	15	0	0	28	38	13	18	66	90
	Men	18	24	0	0	72	74	18	19	600	813
Ethnicity	Kinh	16	17	0	0	28	36	13	18	66	69
	Cotu	17	23	0	0	72	74	18	19	600	834
Age group	<45	18	24	0	0	72	71	19	18	298	560
	>45	16	20	0	0	62	74	17	21	368	343
Health status	Not	17	23	0	0	72	74	18	20	631	859
	Disabled	11	15	0	0	18	24	10	9	34	44
Social status	Not	14	22	0	0	50	74	14	19	524	753
	leader	47	25	7	0	72	48	35	20	141	150
All households		*17*	*23*	*0*	*0*	*72*	*74*	*18*	*19*	*665*	*903*

To reflect relative inequality, we used Gini and Robin Hood indices to examine the distribution of income among households before displacement and after resettlement. It revealed that Gini and Robin Hood indices were high before displacement, 0.48 and 0.38 respectively. These indicated that income distribution among households was inequitably distributed in the former village. Although it decreased in 2009 as Gini and Robin Hood indices declined to 0.37 and 0.26, correspondingly, the total of income of village in 2009 reduced by 84%, compared with before resettlement. After that, the inequality began increasing because Gini and Robin Hood indices went up to 0.42 and 0.33 in 2011 and 0.53 and 0.36 in 2013 (See Figure 5-5). It was much higher than the Gini coefficient of Vietnam, which was 0.43, which is considered as the alarming threshold of inequality (GSO, 2013). Precisely, the poorest 75% of the Bo Hon village had only 41% of total income, whereas the richest 25% accounted for 59% of total income.

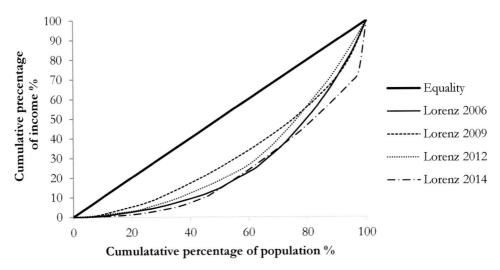

Figure 5-5 Lorenz curve and income distribution before and after displacement of Bồ Hòn village (n = 40 households). *Source: Author's research*

5.7   Conclusions and discussion

Our findings indicated that Bo Hon village made a faster recovery than other communities displaced by hydropower dams in Vietnam (Cao, 2003; CODE, 2010; Ty, et al., 2013; VUSTA, 2008). After 8 years, their income was slightly higher than before displacement and many vulnerable households escaped from poverty. Most became less vulnerable or more resilient to food insecurity and marginalization thanks to a significant enhancement of income. We found that two important conditions contributed to the considerable rehabilitation of the village after resettlement. Firstly, the convenient location of the new village enabled them to access better infrastructure, education, health care, water and electricity, sanitation, and especially labour and agriculture product markets. The village was situated in the strongly developed Acacia forest plantation region where jobs were available to displaced people. Therefore, wage labour income became the most important part of total income after resettlement. The agriculture product demand of urban market enabled them to cultivate cash crops to earn daily income. Although it was small income, it was important for women because they used it to exchange products with retailers from Hue City every day. They also lived close to Kinh people in neighbouring villages, so that they learnt from them how to initiate new livelihood activities migration to cities and industrial zones, as an example. As well, tailoring skills acquired from Kinh people allowed for migration to cities to work in textile and garment factories. As a result, the remittance from worker migration helped to significantly overcome food insecurity.

The second factor was that displaced households received strong support of local authorities who allowed them to reclaim unused uplands for Acacia forest plantation right after resettlement. In addition, the assistance of local and international NGOs enhanced their land security, financial, and technical capacity to invest in Acacia forest plantation. As a result, they had more forestland than before, and the resulting income became the second largest percentage of total income. Thus, we can see that displaced households developed *both land-based livelihoods and agricultural products and labour market-oriented livelihood model* that secured and improved significantly their livelihood outcomes after resettlement.

Thus, the longitudinal study provided greater insights to the resilience process in the case of disturbance of dam-induced displacement and resettlement, especially livelihood resilience. Displaced households applied a mix of livelihood strategies after resettlement, including compensation money spending for survival; restoration of traditional livelihoods and acceptance of new culture for coping; crop diversification; land reclamation for forest plantation; wage labour and migration for long-term adaptation or improved resilience. The case study also showed that displaced households have quite good capacity to self-organize when livelihood opportunities are available, including accessible land, labour market access, good infrastructure and public services, access to urban areas, and good interaction with outsiders, even if they had little support or benefit sharing from hydropower developers after resettlement.

Policy implications

Lessons learnt from the case study and literature review allow us to come up with the implication that relocating a displaced community to the vicinity of urban areas is another positive alternative to resettlement in mountainous areas under the land scarcity and poor natural resources, even though in cities displaced people often encounters the problem of employment capacities that prevent accessing to the competitive employment sphere (Wilmsen, 2011). However, the economy of a region should continue to grow to ensure available opportunities for farm and non-farm employments. In turn, they will be able to exchange position for future generations to repeat this livelihood pathway because, as Scudder (1997) emphasized, successful resettlement takes time; at minimum it should be implemented as a two-generation process. However, to strengthen livelihood sustainability and improvement, there is a necessity to have a stronger commitment by hydropower investors and a firm political will of government, from central to local levels, in supporting displaced people to develop agriculture and forestry in the resettlement sites and allow for migration work in cities. The resettlement program of Three Georges Dam in Zigui, Hubei province, China might be a good example (Wilmsen, 2011) from which to learn and work better for Vietnam and other developing countries in similar conditions.

However, it is quite different in Vietnam because displaced people only migrate temporarily to work in cities or industrial zones, so the support program should be designed in the following way. Firstly, the government should establish a Development Assistance Fund at a central level to gather a certain percentage of profit from hydropower plants or collect funding from other sponsors and reallocate to a provincial development assistance fund. Each province should find their partnerships, for example textile and garment factories in their own cities or other cities, to establish an employment support program. Such a program should provide professional trainings, recruiting, and employment to relocated people.

Another reason textile and garment companies should provide more favourable conditions and priorities for displaced people is because displaced people have been relocated to allow for electricity production that secured the strong development of the textile and garment industry in Vietnam. Still, it is very necessary to involve many other stakeholders to develop innovative programs to improve the resettlement policies, such as INGOs and civil society organizations, to facilitate the government to formulate good resettlement policies and put them into practice. Further studies should analyse the role and participation of important stakeholders related to hydropower dam projects to build a partnership network that help displaced people have sustainable livelihood and improved living standards.

Vietnam promulgated the new Land Law in 2013 for resolving problems of land acquisition for public interests and socio-economic development projects, but they are still low in developing long-term compensation packages and employment opportunities for resettled people. Therefore, it is urgent for Vietnam to issue additional programs besides traditional compensation measures to secure the sustainable development of displaced communities.

References

Adger, W. N. (2000). Social and ecological resilience: Are they related? *Progress in Human Geography, 24*(3), 347-364. doi:10.1191/030913200701540465

Adger, W. N. (2006). Vulnerability. *Global Environmental Change, 16*(3), 268-281. doi:http://dx.doi.org/10.1016/j.gloenvcha.2006.02.006

Agnes, R. D., Solle, M. S., Said, A., & Fujikura, R. (2009). Effects of construction of the Bili-Bili dam (Indonesia) on living conditions of former residents and their patterns of resettlement and return. *International Journal of Water Resources Development, 25*(3), 467-477. doi:10.1080/07900620902965186

Bartolome, L. J., de Wet, C., Mander, H., & Nagaraj, V. K. (2000). *Displacement, resettlement, rehabilitation, reparation, and development. WCD thematic review I.3 prepared as an input to the world commission on dams.* World Commission on Dams: Cape Town.

Berkes, F., J. Colding, & and C. Folke. (2003). Navigating social-ecological systems: Building resilience for complexity and change. . *Ecology and Society, 9*(1)

Binh Dien Hydropower Joint Stock Company. (2011). Binh dien hydropower. 05/03, 2014, Retrieved from http://www.thuydienbinhdien.com.vn/en/index.php?option=com_content&view=article&id =47&Itemid=213

Binh Thanh commune. (2008). *Report on displacement and resettlement in Binh Dien hydropower plant construction (in vietnamese).* Binh Thanh commune: Thua Thien Hue province, Vietnam.

Blaikie, P., Cannon, T., Davis, I., & Wisner, B. (1994). *At risk: Natural hazards, people's vulnerability, and disasters.* London: Routledge.

Bui, T. M. H., & Schreinemachers, P. (2011a). Resettling farm households in northwestern Vietnam: Livelihood change and adaptation. *International Journal of Water Resources Development, 27*(4), 769-785.

Bui, T. M. H., & Schreinemachers, P. (2011b). Resettling farm households in northwestern Vietnam: Livelihood change and adaptation. *International Journal of Water Resources Development, 27*(4), 769-785. doi:10.1080/07900627.2011.593116

Bui, T. M. H., & Schreinemachers, P. (2013). Hydropower development in Vietnam: Involuntary resettlement and factors enabling rehabilitation. *Land use Policy, 31*(0), 536-544. doi:10.1016/j.landusepol.2012.08.015

Cao, T. T. Y. (2003). *Towards sustainability of Vietnams' large dams resettlement in hydropower projects* (MSc thesis). Royal Institute of Technology, Sweden.

Carpenter, S., Walker, B., Anderies, J. M., & Abel, N. (2001). From metaphor to measurement: Resilience of what to what? *Ecosystems, 4*(8), 765-781. doi:10.1007/s10021-001-0045-9

Centre for resource and environmental survey. (2013). *Land survey in 2013*. Centre for resource and environmental survey: Hue, Vietnam.

Cernea, M. (1997). The risks and reconstruction model for resettling displaced populations. *World Development, 25*(10), 1569-1587. Retrieved from http://search.proquest.com/docview/61524231?accountid=14772

Cernea, M. (2003). For a new economics of resettlement: A sociological critique of the compensation principle. *International Social Science Journal, 55*(175), 37-45. doi:10.1111/1468-2451.5501004

Cernea, M. (2009). Introduction: Resettlement – an enduring issue in development. *The Asia Pacific Journal of Anthropology, 10*(4), 263-265. doi:10.1080/14442210903079756

Cernea, M. M. (1995). Understanding and preventing impoverishment from displacement: Reflections on the state of knowledge. *Social Action, 45*(3), 261-276. Retrieved from http://search.proquest.com/docview/61579979?accountid=14772

Cernea, M. M. (2000). Some thoughts on research priorities. *Eastern Anthropologist, 53*(1-2), 3-12. Retrieved from http://search.proquest.com/docview/60401419?accountid=14772

Cernea, M. M., Mathur, H. M., & McDowell, C. (2009). Can compensation prevent impoverishment? reforming resettlement through investments and benefit-sharing. *Journal of Refugee Studies, 22*(1), 130-132. Retrieved from http://search.proquest.com/docview/37230762?accountid=14772

Chambers, R. (1995). *Poverty and livelihoods: Whose reality counts?*. IDS Discussion Paper 347. IDS: UK.

Chambers, R. and Conway, G.R. (1992). *Sustainable rural livelihoods: Practical concepts for the 21st century*. IDS Discussion Paper 296: UK.

CODE. (2010). *Displacement, resettlement, living stability and environmental and resources protection in hydropower dam projects*. CODE: Ha Noi, Vietnam.

126

Cotula, L., Vermeulen, S., Leonard, R., & and Keeley, J. (2009). *Land grab or development opportunity? agricultural investment and international land deals in Africa.* IIED/FAO/: London and Rome.

Cutter, S. L., Barnes, L., Berry, M., Burton, C., Evans, E., Tate, E., & Webb, J. (2008). A place-based model for understanding community resilience to natural disasters. *Global Environmental Change, 18*(4), 598-606. doi:http://dx.doi.org/10.1016/j.gloenvcha.2008.07.013

Dao, N. (2010). Dam development in Vietnam: The evolution of dam-induced resettlement policy. *Water Alternatives, 3*(2), 324-340.

Dao, N. (2011). Damming rivers in Vietnam: A lesson learned in the Tay Bac region. *Journal of Vietnamese Studies, 6*(2), 106-140. Retrieved from http://search.proquest.com/docview/902084590?accountid=14772

Davies, S., & Hossain, N. (1997). *Livelihood adaptation, public action and civil society: a review of the literature.* IDS Working Paper 57. IDS: UK.

De Haan, L. J. (2000). Globalization, localization and sustainable livelihood. *Sociologia Ruralis, 40*(3), 339-365. doi:10.1111/1467-9523.00152

De Haan, L., & Zoomers, A. (2005). Exploring the frontier of livelihoods research. *Development and Change, 36*(1), 27-47. doi:10.1111/j.0012-155X.2005.00401.x

Department of Co-operatives and rural development. (2007). *Displacement and resettlement policies for national programs in the mountainous areas and ethnic minority groups - problems needs to be addressed.* Ministry of agriculture and rural development: Ha Noi.

DFID. (1999). *Sustainable livelihoods guidance sheets.* DFID: UK.

DFID. (2001). *Sustainable livelihoods guidance sheets* Department for International Development: London.

Downing, T. E., (1996). Mitigating social impoverishment when people are involuntarily displaced In C. McDowell (Ed.), *Understanding impoverishment* Providence, Oxford: Berghahn Books.

Dwivedi, R. (2002). Models and methods in development? Induced displacement (review article). *Development and Change, 33*(4), 709-732. doi:10.1111/1467-7660.00276

Ellis, F. (2000). *Rural livelihoods and diversity in developing countries.* Oxford University Press.: New York.

Fernandes, W., (2008). India's forced displacement policy and practice: Is compensation up to its functions? In M. M. Cernea & H. M. Mathur (Ed.), *Can compensation prevent impoverishment? reforming resettlement through investments and benefit-sharing,* (pp. 180-207). New Delhi: Oxford University Press,.

Folke, C., S. Carpenter, T. Elmqvist, L. Gunderson, C.S. Holling, & and B. Walker. (2002). Resilience and sustainable development: Building adaptive capacity in a world of. transformations. *AMBIO, 31*, 437-440.

Fujikura, R., & Nakayama, M. (2013). The long-term impacts of resettlement programmes resulting from dam construction projects in Indonesia, Japan, Laos, Sri Lanka and turkey: A comparison of land-for-land and cash compensation schemes. *International Journal of Water Resources Development, 29*(1), 4-13. doi:10.1080/07900627.2012.741032

Gebre, Y. (2003). Resettlement and the unnoticed losers: Impoverishment disasters among the gumz in ethiopia. *Human Organization, 62*(1), 50-61.

Ha, H. T. (2013). *Thực trạng và định hướng phát triển rừng keo lai tại tỉnh thừa thiên huế (the status and planning for acacia development in Thua Thien Hue province).*

Institute of strategy and policy on resources and environment. (2009). *The status of resettlement programs in the hydropower dams in Vietnam.* Ministry of natural resources and environment: Ha Noi.

Jayewardene, R., (2008). Can displacement be turned into development by compensation alone?. In M. M. Cernea & H. M. Mathur, (Ed.), *Can compensation prevent impoverishment? reforming resettlement through investments and benefit-sharing* (pp. 233-259). New Delhi: Oxford University Press.

Karimi, S., & Taifur, W. D. (2013a). Resettlement and development: A survey of two of Indonesia's Koto Panjang resettlement villages. *International Journal of Water Resources Development, 29*(1), 35-49. doi:10.1080/07900627.2012.739539

Luers, A. L., Lobell, D. B., Sklar, L. S., Addams, C. L., & Matson, P. A. (2003). A method for quantifying vulnerability, applied to the agricultural system of the yaqui valley, mexico. *Global Environmental Change, 13*(4), 255-267.

Maio, F. G. D. (2007). Income inequality measures. *J Epidemiol Community Health, 61*(10), 849-852. doi:doi: 10.1136/jech.2006.052969

Manatunge, J., & Takesada, N. (2013). Long-term perceptions of project-affected persons: A case study of the Kotmale dam in Sri Lanka. *International Journal of Water Resources Development, 29*(1), 87-100. doi:10.1080/07900627.2012.738496

Mathur, H. M. (1999). Restoring incomes and livelihoods of project-affected people: Issues in resettlement planning. *Scandinavian Journal of Development Alternatives and Area Studies, 18*(4), 51-75.

Mathur, H. M. (2013). *Displacement and resettlement in India: The human cost of development.* USA and Canada: Routledge.

Mcdonald, B., Webber, M., & Yuefang, D. (2008). Involuntary resettlement as an opportunity for development: The case of urban resettlers of the three gorges project, china. *Journal of Refugee Studies, 21*(1), 82-102. doi:10.1093/jrs/fem052

McDowell, C. (Ed.). (1996). *Understanding impoverishment: The consequences of development-induced displacement.* UK: Berghahn Books.

McDowell, C. (2002). Involuntary resettlement, impoverishment risks, and sustainable livelihoods. *The Australasian Journal of Disaster and Trauma Studies, 2*

Michelle, M. N. (2011). *Power to people: Hydropower resettlement compensation, democratic land governance and participation in Nong Dan village, central Vietnam.* (MSc Thesis). Universiteit Utrecht.

MOIT. (2013). *Kết quả rà soát quy hoạch, đầu tư xây dựng và vận hành các dự án thủy điện trên cả nước.*

MOIT, Ministry of Industry and Trade. (2013). *Result of hydropower dam construction survey nationwide.* Ministry of Industry and Trade: Ha Noi, Vietnam.

Nakayama, M., Gunawan, B., Yoshida, T., & Asaeda, T. (1999). Resettlement issues of Cirata dam project: A post-project review. *International Journal of Water Resources Development, 15*(4), 443-458. doi:10.1080/07900629948709

Parasuraman, S. (1999). *The development dilemma: Displacement in india.* St. Martin's Press.

Payne, P., & Lipton, M. (1994). *How third world rural households adapt to dietary energy stress.* International Food Policy Research Institute: Washington.

Picciotto, R., van Wicklin, W., & Rice, E. (Eds.). (2001). *Involuntary resettlement: Comparative perspectives.* New Brunswick and London: Transaction Publishers.

Roy, A. (1999). *The greater common good.* New Delhi: India Book Distributors.

Sapountzaki, K. (2007). Social resilience to environmental risks: A mechanism of vulnerability transfer?. *Management of Environmental Quality: An International Journal, 18*(3), 274-296.

Satyanarayan, G. (1999). *Development: Displacement and rehabilitation.* Jaipur: Rawat Publications.

Scoones, I. (1998a). *Sustainable rural livelihoods: A framework for analysis.* IDS Working Paper 72. Institute of Development Studies: Brighton.

Scoones, I. (1998b). *Sustainable rural livelihoods: A framework for analysis.* IDS Working. Institute of Development Studies: Brighton.

Scoones, I., & Wolmer, W. (2002). *pathways of change in africa. crops, livestock and livelihoods in mali, ethiopia and zimbabwe .* James Currey: Oxford.

Scoones, I. (2009). Livelihoods perspectives and rural development. *The Journal of Peasant Studies, 36*(1), 171-196. doi:10.1080/03066150902820503

Scudder, T., (1997). Social impacts of large dam projects In Dorcey, T., Steiner, A., Acreman, M., Orlando, B. (Ed.), *Large dams. learning from the past, looking at the future* (pp. 41-68). Washington, DC,: World Bank.

Scudder, T. (2005). *The future of large dams: Dealing with social, environmental, instituational, and political costs.* UK and USA: Earthscan.

Sisinggih, D., Wahyuni, S., & Juwono, P. T. (2013). The resettlement programme of the wonorejo dam project in Tulungagung, Indonesia: The perceptions of former residents. *International Journal of Water Resources Development, 29*(1), 14-24. doi:10.1080/07900627.2012.743432

Souksavath, B., & Nakayama, M. (2013). Reconstruction of the livelihood of resettlers from the nam theun 2 hydropower project in Laos. *International Journal of Water Resources Development, 29*(1), 71-86. doi:10.1080/07900627.2012.738792

Stanley, J. (2004).
 Development-induced displacement and resettlement.

Tan, Y., & Wang, Y. Q. (2003). Rural resettlement and land compensation in flooded areas: The case of the three gorges project, china. *Asia Pacific Viewpoint, 44*(1), 35-50. doi:10.1111/1467-8373.t01-1-00182

The General Statistics Office of Vietnam. (2013). *Statistical handbook of Vietnam.* The General Statistics Office of Vietnam: Ha Noi, Vietnam.

Thua Thien Hue People's Commitee. (2013). Socio-economic development data of Thua Thien Hue province in 2013. 05/22, 2014, Retrieved from http://www1.thuathienhue.gov.vn/portal_es/Views/Article.aspx?CMID=28&TLID=224

Tien, D. (2013). Năm 2013, thêm 200.000 công nhân ngành dệt-may sẽ có việc làm (more than 200,000 workers in textiles industry will have jobs in 2013). 05/11, 2014, Retrieved from http://tamlongvang.laodong.com.vn/cong-doan/nam-2013-them-200000-cong-nhan-nganh-detmay-se-co-viec-lam-99290.bld

Ty, P. H., (2008). Diplacement and resettlement due to Binh Dien hydropower construction: How is the impoverishment risks managed? In Le Duc Ngoan (Ed.), Hue city: The publisher of Hue University.

Ty, P. H., Phuc, N. Q., & and Westen, G. v. (2014). Vietnam in the debate on land grabbing: Conversion of agricultural land for urban expansion and hydropower development. In Mayke Kaag and Annelies Zoomers (Ed.), *The global land grab: Behind the hype* (pp. 135-151). London and New York: ZED.

Ty, P. H., Hien, N. T., Hien Pham Dinh, & and Anh, D. C. (2010). *Ảnh hưởng của việc di dân tái định cư để xây dựng thuỷ điện bình điền đến hoạt động sinh kế của người dân tộc thiểu số thôn bồ hòn (impacts of displacement and resettlement due to binh dien hydropower dam construction on ethnic minority groups in thua thien hue province)*. Centre for Rural Development: Hue, Vietnam.

Ty, P. H., Van Westen, A. C. M., & Zoomers, A. (2013). Compensation and resettlement policies after compulsory land acquisition for hydropower development in vietnam: Policy and practice. *Land, 2*(4), 678-704. doi:10.3390/land2040678

VUSTA. (2006). *A work in progress: Study on the impacts of vietnam's son la hydropower project
*. Vietnam Union of Science and Technology: Ha Noi, Vietnam.

VUSTA. (2008). *Follow-up study on impacts of resettlement of son la hydropower plant.* Vietnam Union of Science and Technology Association: Ha Noi, Vietnam.

WCD. (2000). *Dams and development, a new framework for decision-making. the report of the world commission on dams* Earthscan Publications Ltd: London/Sterling, VA.

Wilmsen, B. (2011). Progress, problems, and prospects of dam-induced displacement and resettlement in china. *China Information, 25*(2), 139-164. doi:10.1177/0920203X11407544

World Bank. (2004). *Involuntary resettlement sources book: Planning and implementation in development projects.* Washington, DC: The International Bank for Reconstruction and Development.

Yoshida, H., Agnes, R. D., Solle, M., & Jayadi, M. (2013). A long-term evaluation of families affected by the Bili-Bili dam development resettlement project in south Sulawesi, Indonesia. *International Journal of Water Resources Development, 29*(1), 50-58. doi:10.1080/07900627.2012.738495

Zoomers, A. (2010). Globalisation and the foreignisation of space: Seven processes driving the current global land grab. *The Journal of Peasant Studies, 37*(2), 429-447. doi:http://dx.doi.org/10.1080/03066151003595325

Zoomers, A., & Albó, X. (2000). Linking livelihood strategies to development. experiences from the bolivian andes. *European Review of Latin American and Caribbean Studies, 69*, 123-124. Retrieved from http://search.proquest.com/docview/38980202?accountid=14772

6 THE ROLES AND INFLUENCES OF NGOS IN GOVERNING HYDROPOWER DEVELOPMENT IN VIETNAM[27]

6.1   Introduction

Literature review shows that the Vietnamese party-state gave little autonomy to Non-Government Organizations (NGOs) between 1975 and 1986. However, the context has changed after the economic reform of Vietnam beginning in 1986. The transition from a command, centralized, and bureaucratic economy to market-oriented economy has made significant transformation of the economic and social conditions and international integration. Following that, social forces have increasingly matured and gathered into an important force in Vietnamese society in recent years, particularly the emergence of Vietnamese non-governmental organizations (VNGOs). According to Bui (2013), approximately 1000 VNGOs exist in Vietnam that focus on research, consultancy, education and health promotion, community development, and poverty reduction. Additionally, over 950 international non-governmental organizations (INGOs) have been created, which have spent more than $2 billion US in 2013, of which 37% was allocated for health care; 17% for economic development; 17% for education; 20% for social services; and 9% for emergency relief. Through these activities, a relationship between NGOs and the party-state has been established. This relationship can be also formed in many contested arenas and social movements, including corruption, agricultural cooperatives, media (Kerkvliet, 2001), bauxite mining, Catholic land conflict, political opposition, and bloggers (Thayer, 2009). These social movements have been expanded to fight for the legitimacy and autonomy of civil society organizations (CSOs) in Vietnam. Previous movements were led by the elites who were once leaders of the party-state, and they tended to be dismissed after problems were addressed (Bui, 2013). However, many NGOs have initiated campaigns in strategic and long-term plans, which are linked to the controversial issues of hydropower dam construction.

In this paper, we argue that the emergence of NGOs in hydropower development has shaped a new political space in Vietnam. Therefore, this paper aims to look at several actions and campaigns initiated by NGOs to strengthen their autonomy and to establish new relations with the party-state of Vietnam in order to examine their roles and influences on hydropower development in Vietnam. This paper will answer several questions: how did NGOs become involved in hydropower development? What actions or programs did they initiate to influence hydropower

---

[27] *This chapter will be submitted as* Ty, P. H., Zoomers, A., & Van Westen, A. C. M.. The roles and influences of NGOs in governing hydropower development in Vietnam. *Journal of Development and Change.*

development decision-making? How did they interact with government? What can we learn from this interaction? To answer these questions, we looked at the engagement of VNGOs and INGOs who initiated campaigns to oppose hydropower dam development or to improve displacement and resettlement outcomes due to hydropower dam construction. Most data and information were collected from reports of NGOs, conferences, independent evaluation reports of NGOs and consultancy institutes, online newspapers, interviews with displaced people, staffs of local authorities, and leaders of NGOs in Vietnam.

The paper is structured as follows. Firstly, following the introduction, an analytical framework will argue as to the relation of the state with civil society organizations (CSOs) and NGOs in the context of single party-state countries. This argument lays a theoretical foundation for analysing the NGOs-state relation in Vietnam. Secondly, the NGOs management system of Vietnam government is presented to analyse the relationship between civil NGOs and the state. Thirdly, the roles and influences of VNGOs and INGOs are described in three case studies. Finally, a discussion and conclusion will anchor the main findings and draw lessons from the involvement of INGOs and VNGOs in hydropower development in Vietnam.

6.2   Analytical framework

NGOs are a part of civil society, and, therefore, it is necessary to delineate exactly the boundary among them. However, it is very difficult to define this with absolute clarity because relationships are often intertwined (Sinh, 2002; Metzger, 1993; Kerkvliet, 2001; Nørlund, 2007). In a broadest sense, civil society has been characterized as a public sphere of social life that excludes government activities (Meidinger, 2001). The term *civil society* is usually applied to categorize individuals, institutions, and organizations that aim at proceeding or conveying a common goal through ideas, actions, and demands on governments (Cohen & Arato, 1992). *State* is regarded as officials and institutions that create, perform, and enforce rules that are envisioned to apply to the whole society and its different components.[28] No society, however, is thoroughly uniform (Kerkvliet, 2011).

---

[28] Kerkvliet (2001) included here both the physical and structural aspects of a state (buildings, offices, army, bureaucrats, government officials, roadways, and so forth), as well as both the ideological dimension and psychological impact of a state. The state includes agencies that keep the wheels of government turning (issue passports, police cities, create new laws, punish violators, mobilize armies, collect taxes, build new buildings); but the state also has purposes, plans, objectives.

133

In Western countries, such boundaries redefined as autonomous areas of individuals, social groups, and organizations that are separated from activities of the state, and there is no direct control of the state (Meidinger, 2001). Activities are organized on the basis of voluntary, self-generating, self-reliance, self-supporting, peaceful coexistence, and to comply with laws and regulations set out as rules for each organization and to be capable of asking the state to involve them in decision-making, while, simultaneously, resisting subordination to the state (Cohen and Arato, 1992; Falk, 1995; Habermas, 1991; Diamond, 1994; Oxhom, 1995). According to liberal democratic theory, to legitimize state authority and to limit state power abuses, a strong and plural civil society is essential. In particular, civil society should been titled to civil and political rights, which include freedom of association. Civil society involvement is necessary to ensure that the state provide an accountable government that is subject to free and fair elections (Diamond, 1994). By directing and handling needs and concerns of various interest groups to the state, civil society is the foundation to ensure legitimacy, accountability, and transparency and to enhance the state's capacity to good governance (Mercer, 2002; Diamond, 1994, p.7). Therefore, civil society and the state are separate but mutually supportive (Baker, 1997).

Following the liberal democratic views on civil society, the existence of civil society in authoritarian states, such as China, Vietnam, and Laos, is very limited. However, alternative ways exist to examine the emergence of civil society in authoritarian states (Heurlin, 2010; Bui, 2013). Ramasamy (2004) noted that Gramsci's theoretical framework is more useful for exploring civil society in Southeast Asia, which is often ignored by liberal perspectives. The biggest difference between liberal theory and Gramsci's is the degree of freedom or autonomy of civil society from the state that a civil society enjoys and, more broadly, in the nature of its relationship to the state. According to Gramsci, civil society is an "arena of contestations" in which the state is engaged in a struggle with other actors to dominate popular ideas, values, and norms (Ramasamy, 2004). Therefore, not only does the state attempt social control through coercive or regulatory means, but it also exercises ideological hegemony by manufacturing cultural and ideological consent within civil society (See Femia, 1981, p. 31-35). Gramsci argued that the state's hegemony over civil society is never complete. Whoever controls the arena of civil society succeeds in manufacturing consent for political domination (Ramasamy, 2004, p. 203). Kerkvliet (2001) suggested a new approach for analysing state-society relations in Vietnam; he investigated the state-civil society relations through contested arenas; for instance, corruption, Catholic land disputes, and political dissents and oppositional bloggers. Thus, according to views of Gramsci and Kerkvliet (2001), the relationship of civil society with authoritarian states has different characteristics compared to that of liberal democratic states.

In civil society, non-governmental or non-profit organizations (NGOs) play an important role. They are usually formally established, managed by employed staff (often urban professionals or foreigners), well-supported (by domestic or international capital), and often relatively large and well-resourced. NGOs are usually created for a multitude of reasons to participate in human imagination and aspirations. They can beset up to support specific aim or goal, such as human rights or to implement programs on the ground, such as disaster relief. They may have members ranging from local to global (Charnovitz, 1997, p. 186). The activities of NGOs influence civil society and the state differently. NGOs may strengthen the state or civil society in this context, but may weaken or destroy them in other cases (Mercer, 2002, p. 12). NGOs strengthen the state through their involvement in improving the efficiency of government services and acting as a strategic partner for the reform-oriented ministries. They fill in gaps in services and supply and connect the state with the grassroots. Simultaneously, NGOs strengthen civil society through oppositional actions, such as demonstrations to political demands and to provide a separate channel from Congress through which disputes can be negotiated and dissipated (Clarke, 1998, p. 211). Also, NGOs usually act to promote democracy and human rights through opening spaces for the participation of citizens and organizations in ways that follow state laws. But, at the same time, they may create a network of citizens to protest for insisting that the state to modify laws and regulations. Currently, many NGOs are operating independently and acting as "watchdog" vis-à-vis the state (Mercer, 2002; Fisher, 1998, p. 16).

However, principles and actions of NGOs also express threats to state power and its legitimacy, and weaken the role of the state in the provision of social services (Fowler, 1991; Tvedt, 1998; Marcussen, 1996). Therefore, authoritarian states apply two strategies to control NGOs: exclusionary and corporatism approaches.

The first strategy aims to curb the growth of NGOs through harassment, including setting out the limits on the fields and locations where NGOs can operate (Bratton, 1989); replacing the work of NGOs by the state institutions or even government-organized NGOs (GONGOs) or umbrella groups, particularly related to development activities and social welfare provision; setting high barriers for registration and denying formal legal protections (Sakai, 2002); placing severe restrictions to thwart NGOs' ability to raise funds domestically and to access foreign funds (Riker, 1995). This strategy was used in Indonesia (1971–1998), the Philippines (1972–1986), China (1949–1978), and Vietnam (1975–1986) (Heurlin, 2010).

For the second strategy, *corporatism*, totalitarian states usually formulate GONGOs and umbrella institutions to control the activities of NGOs. These organizations play a central role in organizing the registration, review, and coordination of activities of NGOs. From this, information on NGOs, including the types, sizes, scopes of activities, locations of operation, and their relationships with other civil society organizations CSOs, is collected, monitored, and reported for state leaders. These institution-building efforts may not only result in better control over co-opted NGOs, but also encourage independent NGOs to actively seek incorporation with the state (Foster, 2001). Thus, GONGOs and umbrella institutions play roles as "watchdogs" over independent NGOs and as fund-raisers from international states and donors (Schmitter, 1974; Lehmbruch & Schmitter, 1982). Furthermore, the state co-opts independent NGOs at three levels. Firstly, the state encourages former and active government officials to participate in NGOs leadership positions. Secondly, the leaders of independent NGOs are considered for appointment to leadership positions of state agencies or invited to join as members of advisory boards or political bodies established by the state (Lehmbruch & Schmitter, 1982; Riker, 1995; Foster, 2001; Ma, 2002). For single-party states, the chance of disaffected cadres mobilizing NGOs to challenge the party is much lower. Single-party states in general tend to co-opt their critics (Geddes, 1999, p. 135). Finally, the state can provide NGOs with land, office, infrastructure, and other favourable conditions for the activities of NGOs (Foster, 2002; Ma, 2005). Authoritarian rulers in China (1978–present), Taiwan (under the KMT until 1987), and Vietnam (1986–present) have all chosen a corporatist strategy to govern NGOs (Heurlin, 2010).

It would be a mistake to think that these exclusionary and corporatist strategies are applied only in totalitarian countries; to some degree, both strategies are used in democracies. However, the two strategies are deployed in different ways. Japan was one of democracies that applied the exclusionary strategy to regulate the establishment of NGOs during 1990s. In this period the Japanese state dominated development policy, which relied upon an independent state bureaucracy and weak parliament (Goodman et al., 1998; Pekkanen, 2006). In contrast, Germany and France chose a corporatist strategy for governing NGOs. Salamon & Anheier (1998) argue that corporatist strategies have emerged in democracies in which the state has been forced or induced by key social elites into making common cause with NGOs. This strategy is employed in order to pre-empt more radical demands for social welfare provision. In this paper, we focus on analysing the relationship between civil society and NGOs within a single-party state, and, therefore, the relationship between civil society and NGOs within democratic states is not a central point to discuss. In what follows, we will look closely at the administration system of Vietnam in controlling NGOs autonomy and influences.

6.3   Governing civil societies/NGOs in Vietnam

Gramsci indicated that civil society in single-party states is an arena of contestations in whichthe state engages to struggle with other actors to place in dominance their ideology, values, and norms. This is particularly true of the plight of Vietnam after independence in 1945 to date. The Communist Party has recognized the importance of ideological hegemony, and, therefore, the party has dominated the ideology of entire Vietnamese people over other social organizations. To unify thought and action cross the country, the party has undertaken the slogan "building the great solidarity for the whole country "by creating a unique political system comprised of three political pillars: the Communist party, the state, and society (Sakata, 2006). Instead of self-establishment, social organizations are established by the party-state as a part of a broader political system. Within this system, the party-state supposes that the political and social regime is unified, and there is no need to have other social organizations outside this system. With this awareness, the Communist Party's determination is to prevent "non-socialist" principles (Marr, 1994, p. 11; Hansson, 2003). Civil society organizations outside of this political system are not officially allowed to operate on the ground. The term *non-governmental organizations* are not regulated by law, instead the term *non-profit organizations* is employed to characterize independent, social organizations whose functions are like that of NGOs'. As a result, this political context has shaped a debated and contested relationship between the party-state and NGOs (Thayer, 2009; Heurlin, 2010; Landau, 2008; Bui, 2013).

In practice, the party-state has employed two administration strategies to control NGOs in Vietnam for different periods. Between 1975 and 1986, the party-state undertook exclusionary strategies to manage NGOs (Heurlin, 2010, p.223; Bui, 2013). NGOs were not allowed to be independent in operation (Landlau, 2008, p.250). During this period, the Party was suspicious of and reluctant to allow the free activities of NGOs that promoted democratization and human rights (Landau, 2008). Since 1986, however, the implementation of a market-oriented economy weakened the Party's grip on society (McCormick, 1999, p. 153; Thayer, 1992, p. 110; Thayer, 1995, p. 39). The party-state has changed their strategy to manage NGOs from exclusionary to corporatist strategy (Heurlin, 2010, p.223; Bui, 2013). To operationalize this strategy, the party-state has established two systems with various umbrella associations to handle VNGOs and INGO separately.

The first system's function is to manage International Non-Government Organizations (INGOs). The state established the Vietnam Union of Friendship Organizations (VUFO), which acts as an umbrella for all INGO are required to register to be legitimate to operate in Vietnam in which the People's Aid Co-ordinating Committee (PACCOM), the specialized and functional body of VUFO, was established in1989 in order to reinforce the partnership between INGOs with government institutions and local authorities and to facilitate INGOs activities in Vietnam, as well as to support local partners to work with INGOs. PACCOM also serves as a focal point to approve and grant license for INGOs. In 2001, the Committee for Foreign NGO Affairs (COMINGO) was formed to assist the Prime Minister in guiding and addressing issues relating to foreign NGOs in Vietnam.

Currently, the organizational structure to control INGO has been built out from central to local levels. Legal documents and guidelines for the registration and management of INGOs have been fully formulated. Decree12/2012/ND-CP, the latest guidance, was issued in 2012to operate the registration and activities of INGOs in Vietnam. This document also stipulates cooperation mechanism between INGOs and government organizations (GOs), in which the state shall encourage and create favourable conditions for INGOs in implementing humanitarian and development activities in Vietnam. But the state prohibits INGOs to carry out political and religious activities that are inconsistent with the interests of Vietnam as determined by the state. INGOs are asked to report both the progress and results of their projects to PACCOM (Nguyen, 2011). In addition, INGOs are required to have local partners in project implementation, including GOs and social-political and professional associations. This type of partnership takes the form of project-based development partnerships. To enhance their political positions and to maintain positive relations with the state, INGOs in Vietnam often choose, or more often are pressured, to collaborate with GOs (Hakkarainen & Katsui, 2009; Nørlund, 2007; Wassermann & Nguyen, 2003).

The second system's function is to manage Vietnamese NGOs. The state has set up umbrella associations for VNGOs to register in terms of different areas. For example, VNGOs working in science and technology must register at the Vietnam Union of Science and Technology Associations (VUSTA) and its sub-unions at provincial levels. Within this umbrella, VNGOs are legitimated to operate according to legal frameworks applying to VUSTA as set out by the Vietnamese government. Recently, the state has opened more spaces for VNGOs by giving them more autonomy and independence. For example, decision 650/QD-TTg (April 2006) of the Prime Minister has given more rights for VNGOs to be autonomous and to self-organize and participate in monitoring development policies of the government and local authorities (Article 6 and 8). In particular, the government issued decision 14/2014/QĐ-TTg (January 2014) to give more rights on *social consultancy, criticism, and inspection* (SCAI) for VUSTA and its members (VNGOs). This is the first time that these terms were legalized in laws, where VNGOs were able to criticize development policies of the party-state. Within these decisions and new laws, social criticism is defined as activities giving comments, evaluation, critiques, and recommendations on the appropriation and feasibility of policies. Social inspection is regarded as actions identifying the scientific and feasible aspects of policies. As a result, forums have been established on websites of political-social organizations and VNGOs to discuss these concepts. For example, a forum on social supervision and criticism— *diễn đàn giám sát và phản biện xã hội*—was created on the website, Vietnam Fatherland Front (See Figure 6-1). Here, a series of discussions on social supervision and criticism have been published by many former leaders of the party-state (such as former President Nguyen Minh Triet), researchers, and journalists. Furthermore, many VNGOs have clearly declared their missions on social supervision and policy advocacy on their websites, such as the missions of Vietnam River Network (VRN), Centre for social research and development (CSRD), Consultative and Research Centre on Natural Resource Management (CORENARM), and Centre for Innovation and Development (GreenID). Thus, in theory, the state has provided space for more tolerance and endorsement for the autonomy and independence of VNGOs and INGOs to participate in making decisions on development policies. Nonetheless, it is more important to see how these policies have been operating in practice through an examination of several cases that involved VNGOs and INGOs influence in the decision-making process of hydropower dam construction and displacement and resettlement in Vietnam.

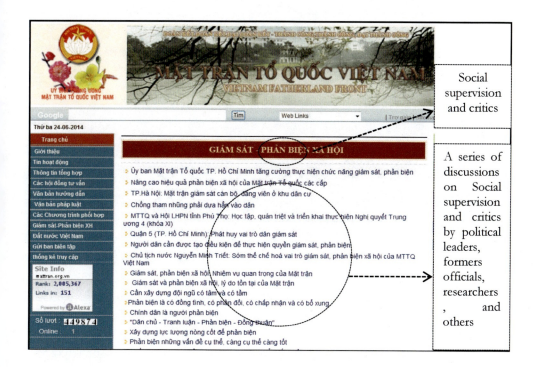

Figure 6-1 A new forum on social supervision and critics on the website of the Vietnam Farther Front. *Source: http://www.mattran.org.vn/Home/GSPBxahoi/gspbxh.htm*

6.4 The roles and influences of NGOs in governing hydropower development

Due to increased concerns on environmental and social problems in hydropower development, many VNGOs and INGOs in Vietnam have been working on three central problems related to hydropower development: lack of transparency when the government and local authorities make decisions on building new dams; insufficient land provision for displaced communities to restore their livelihoods after resettlement; and lack of mechanism to share benefits from producing hydroelectricity for displaced communities. Each problem has been addressed by different actions and campaigns that shaped the relationship between NGOs and the state on the ground. The following case studies were used to characterize how NGOs initiated their actions and influenced the decision-making process of government and local authorities.

Case study 1: Anti-dam movement initiated by VNGOs: Chapter 2 has shown that the hydropower development plan of Vietnam was not opposed by civil society organizations because this development was necessary for the sake of electricity security and economic development. Hydropower was the leading motto because it was considered as a clean and renewable energy, and Vietnam has abundant hydropower potential. Everyone supported and even voluntarily donated their homes and land for hydropower development without asking the state and hydroelectric plants to compensate, as long as the country would have sufficient electricity for development. The opening day of new hydropower plants was once a great festival of the people nationwide and in the project areas. In the late 1980s and early 1990s, postage stamps were designed with Hoa Binh hydropower icons and medals awarded by the state to demonstrate the hope of successful industrialization and modernization (Dao, 2010; Dao, 2012). However, after 2005, the perception of hydropower has strongly shifted. The pride and excitement has gone, and instead is anxiety when hearing of the inauguration of new hydropower plants. This anxiety occurred from a series of tremendous consequences to environment and society at an alarming level. Individual and collective appeals have rapidly increased due to the rapidity of hydropower dam development. For example, 62% of households affected by the Yaly, Ban Ve, and Tuyen Quang dams said that they sent appeals to different levels of government, of which 18.2% was at the community level, 16.8% at the commune level, 29.5% at the district level, and 26.7% at the province level and 8.95 at the central government level (CODE, 2010). Most of the resistance of local communities has been initiated to ask for higher compensation for property losses due to displacement and resettlement. They do not considerably influence the government to change their decisions on building hydropower development.

Only after the Vietnam River Network (VNR) was established in 2005 did this network gather together, on a national scale, many VNGOs and social and environmental activists to oppose hydropower development and to stop new dam construction. VRN is recognized as a growing network of VNGOs, academics, community-based organizations (CBOs), and environmental and social activists working to encourage the sustainable development of Vietnam's rivers. Currently, VRN has 300 individual members and 20 local VNGOs who registered with VUSTA. Staffs of VNGOs are well educated and most are trained abroad. VNGOs also employ former government experts who once worked in the irrigation and hydropower sector, and, therefore, they have a good understanding of hydropower construction and legal framework of Vietnam (Author's interview with leaders of VRN, 2014). VRN is also a partner of International Rivers Network (IRN), a leading international NGO in the fight against dam building worldwide (Biswas, 2012). Therefore, the vision and action approach of VRN is influenced by the risks and rights-based approach proposed by IRN and World Commission on Dams (WDC) in 2000. Furthermore, VRN is a pioneering institution in carrying out independent monitoring, supervision, and evaluation of hydropower dam construction developed by the state, private companies, and development banks, such as ADB and WB (Interview with the coordinator of VRN in Hue city, 2013). They have also organized campaigns to stop building the Xayaburi dam on the Mekong watershed in Northern Laos (Author's interview with leaders of VRN, 2014).

In 2012, VRN initiated an anti-dam movement against the construction of Dong Nai 6 and 6A hydropower dams invested in by a private company, Hoang Long Group, in Dong Nai province. These two dams were approved by the government for investment in 2010, but government leaders, local authorities, and hydropower investors were confronted by widespread opposition of a VNGO alliance to their hydropower development plan. Although the government had had experience with nationwide opposition by elite groups to development plans, such as bauxite mining and forest land lease to foreigners, the government had no experience with resistance led by a network of VNGOs whose efforts were to stop the building of new hydropower dams. This was the first time VNGOs challenged the competency of government to approve large-scale infrastructure development projects. VNR asked the government to stop granting investment license to Dong Nai hydropower 6 and 6A, so VNR could re-evaluated the environmental impact assessment (EIA) reports prepared by the investor to disclose potential impacts of hydropower dams to the public.EIA reports prepared by investors had been approved by the government or local authorities beforehand, but the government did not test the accuracy of those reports.

However, according to a VRN assessment in 2012, information and data on negative impacts of these dams were not correct because the investor ignored several severe consequences to the environment and local society where Dong Nai 6 and 6A were proposed to be built. While the investor estimated that loss of forestland and biodiversity was negligible, VRN revealed considerable losses of rich forests and biodiversity values in many protected areas, including Cat Tien National Parks, UNESCO-recognized Dong Nai Biosphere Reserve, Bau Sau Ramsar Site, and Special Dong Nai National Relic Site. Additionally, VRN was critical, saying that the investors violated many laws. For example, two hydropower projects could acquire 50 ha of land in the core zone of Cat Tien National Park, but the investor did not ask the National Assembly for approval. Further, the investors also violated the Law on Biodiversity, Law on Cultural Heritage, Law on Forest Protection and Development, and Law on Environmental Protection, Laws on Wetland Protection and Development, and many other international conventions on biodiversity and biosphere protection signed by the Vietnamese government. As a result, negative impacts and law violations were quickly transmitted to the populace and state leaders through different channels. VNR organized information sharing conferences with local authorities, affected nature reserves, sub-unions of VUSTA in Dong Nai province, environmental and social activists, and government officials. Also, they propagated information on the websites of many VNGOs and mass media. Their EIA reports were also sent to the Department of Appraisal and Evaluation Committees on EIA reports and the National Assembly Committee on science and technology to request the cancelation of Dong Nai 6 and 6A hydropower plans (VNR, 2013).

With public pressure, hydropower has become a controversial issue in political forums, especially at meetings of National Assembly and Provincial Councils. The accountability of government agencies was questioned and criticised by National Assembly delegates, in which ministers of the Ministry of Industry and Trade (MOIT) and the Ministry of Natural Resources and Environment (MONRE) were asked to explain their responsibilities at many meetings of the National Assembly of Vietnam in 2012. After that, the government asked the National Committee on Environmental Impact Assessment and Approval (NCEIAA) to re-evaluate impacts of two projects. In 2013, the government invited a representative of VRN to participate in NCEIAA and requested the investors organize a press conference to clarify their projects to journalists and environmental and social activists. After the conference, the government decided to stop the construction of Dong Nai 6 and 6A hydropower dams. More importantly, in 2013, the government held a review of all hydropower projects. After review, the government decided to remove 338 hydropower projects, including 2 hydropower terrace projects and 336 small hydropower projects. The government also eliminated 169 potential sites (MOIT, 2013). Furthermore, the government directed the Ministry of Agriculture and Rural Development (MARD) to establish a reforestation project to replace the 76,000 ha area for the loss of forest land due to hydropower, mining, road construction, irrigation, industry, resettlement, and rural infrastructure (Vietnam Administration of Forestry, 2014).

The strong anti-dam movement has also influenced the investment of many private investors, many of which have recently stopped building hydropower dams. After the case of Dong Nai hydropower dam 6 and 6A, few private companies registered with the government and local authorities to build hydropower dams; whereas, there are 169 hydropower projects planned by the government and local authorities nationwide (MOIT, 2013). Furthermore, in 2013, many private companies withdrew their investment from hydropower development. For instance, Hoang Anh Gia Lai Group (HAGL) sold 6 hydropower projects, and Nam Trung Group transferred a proportion of their capital from the investment to Dong Nai 2, Krong No 2, and Krong No 3 hydropower projects. According to the Deputy Secretary of Vietnam Energy Association, the loss of investors from hydropower projects is inevitable because building hydropower dams is very risky at present. He also emphasized that most private companies invested in small hydropower projects not only for the production of electricity but also for exploitation of rich forests in the watershed of reservoirs. Therefore, many projects were proposed near natural forests. After logging forests, most companies did not want to continue constructing and operating hydropower dams (Dong, 2013).

According to the leader of VRN, the success of anti-dam movement was the first victory of the coalition of VNGOs. This victory also showed that the strong alliance of VNGOs, media, and environmental and social activist had strengthened their voices and autonomy. VNGOs have become more powerful and influential to the decision-making process of government and local authorities. This power can be seen as a brake to mitigate the supreme power of the single party-state of Vietnam, and, therefore, help avoid the adverse consequences caused by hydroelectric development. The growth of VNGOs is fortunate; otherwise, poor governance measures of government and local authorities on hydropower development would not be challenged, especially plans for small and medium-sized hydropower dam construction authorized by provincial-level authority and private companies.

The intervention of VNGOs on Dong Nai hydropower projects can be also viewed as a test for the relationship between VNGOs and the state on the ground. The method that VNGOs implemented their campaigns did not create the confrontation atmosphere between VNGOs and the party-state. VNGOs understand the legal system and political context of Vietnam very well, and, therefore, they often create a constructive environment to work with government and local authorities, as well as investors. Instead of depending on the elite network to criticize or oppose the development policies of the party-state, VNGOs lead their campaigns themselves in well-organized methods. From their campaigns, VNGOs strengthen their autonomy and roles in the political system of Vietnam, one that was once restrictive and repressive. The findings also show that government has become more tolerant to existing of VNGOs. Government also has applied the corporatist strategy to co-opt the leaders of VNGOs into the government system. The reaction of government to the anti-dam movement is also a chance for government to increase the trust of entire people to the legitimacy of the party-state.

Case 2: Enhancing the responsiveness of local authorities in allocating land for displaced people led by local VNGOs. In this case, we investigated the roles of two local VNGOs, including the Centre for Social Research and Development (CSRD) and the Consultative and Research Centre on Natural Resources Management (CORENARM), who promoted land allocation programs to communities displaced by dam construction in Thua Thien Hue province, Central Vietnam. Both VNGOs are officially recognized as local NGOs in Thua Thien Hue province. CRSD aims to become a leading NGO, contributing towards government policy-making and influencing community attitudes towards the environment and community development. CORENAM focuses on delivering consultancy services on natural resource management and community development. Both NGOs are financially and administratively independent. They often receive funding from INGOs and intergovernmental organizations. Most staff graduated from overseas universities, and, therefore, they are strongly influenced by the risks and rights-based approach. They usually have good English language skills. Leaders of CRSD are mainly women. Both NGOsare members of VRN, and, therefore, participated in the anti-dam movement against Dong Nai hydropower dam 6 and 6A (Authors' interview with leaders of CRSD).

At local level, CRSD and CORENAM organized actions to facilitate local authorities to solve the problem of a shortage of productive land for displaced people. This is the most challenging issue for most displaced people, who receive much less land for farming as promised by local authorities and investors than promised when persuaded to accept resettlement. [As reported in the previous chapter, inadequate land provision has hampered the reconstruction of traditional livelihoods for displaced people (Ty et al., 2013; Le, 2011).

Between 2010 and 2012, CRSD and CORENAM implemented a project in eight ethnic-minority villages of two communes in Thua Thien Hue province. These villages were resettled due to construction of an irrigation dam and the Binh Dien hydropower dam. The project was funded by ICCO, a Dutch government-funded INGO. The project aimed to help displaced villages access forest land, increase farming skills, and increase gender awareness among displaced people. CSRD and CORENAM negotiated with the district peoples' committee, the provincial Department of Natural Resources and Environment (DoNRE), and Forest Protection Management Unit (FPMU) to allocate 169 ha of unused land to resettled households in four resettlement villages for planting indigenous bamboo and other trees. Of which, 91 ha of forest were allocated to individual households in Hong Tien commune with land use rights certificates (Red Book), and 78 ha of protected forest were allocated to three communities in Binh Thanh commune.

Furthermore, CRSD and CORENAM organized workshops on land laws and land use rights for residents. They provided resettlers with trainings on communication and negotiation skills, raised awareness of gender issues and human rights, and provided resettlers with a forum for discussing post-resettlement problems. They held training courses on sustainable land use, veterinary skills, introduced new crops, and helped resettlers analyse value chains for agro-forestry. Furthermore, they trained local officials in land use planning and land allocation processes. According to Binh Thanh commune officials, the training increased their awareness of residents' needs and willingness to collaborate on development initiatives. CRSD and CORENAM believed that their programs were useful to promote grassroots democracy policy and to put good policies into practice (Authors' interview, 2013).

According to the leaders of local VNGOs, their activities received significant support from local authorities, and, therefore, they could implement successful forest land reallocation programs to displaced communities. Additionally, staffs of local NGOs asserted that local VNGOs are more effective than INGOs in advocating for affected people and working with local officials because they have a greater understanding of the locally administrative context and residents' needs than INGOs (Authors' interview, 2013). They are well-versed in the exigencies of negotiations with the FMPU over land reallocation, working patiently to persuade the board to cede the forest land under its control. They maintain longstanding ties with local governments, and they often bring together representatives of resettled and downstream communities, as well as local VNGOs and local officials to exchange experiences and strategies in workshops. Although they lack the deep pockets of many INGOs, they are experienced in using small budgets effectively. However, the need for extensive negotiations to achieve even limited land reallocation underscores the challenges facing large-scale replication, and local VNGOs success in effecting reallocation of land for resettled villagers by FMPU officials is often contingent on the funding and technical expertise VNGOs can provide.

This case shows that local VNGOs actively advocate for land allocation to resettlers and provide many consultancy services to local authorities and resettled communities. Also, the intervention of local VNGOs challenged the governance performance of local authorities on land allocation for displaced people after resettlement, especially the lack of responsiveness when it comes to providing insufficient land to displaced communities after resettlement. Local VNGOs strengthened their autonomy within the local government system through formulating good partnership with local officials. As a result, their roles have been widely recognized by local leaders, and they have made significant influence on improving the outcomes of displacement and resettlement due to hydropower dam construction. Today, local VNGOs play a role as representatives for vulnerable communities to protect their legitimate rights and interests when it comes to hydropower development in Vietnam.

Case 3: Initiating a new benefit sharing mechanism for displaced people by INGOs. In Vietnam, there is no official benefit-sharing policy related to using revenues from hydropower development to help resettled people restore their livelihood. Displaced people only receive cash and land compensation for their losses, assistance for displacement and resettlement, and supports for restoration of livelihoods (mainly agricultural), and job replacement within 1 to 2 years after resettlement, which are not effective enough to help them rebuild their lives and livelihoods. Most become poorer after resettlement because they do not have sufficient jobs and income because the responsibility of government, local authorities, and hydropower authority for improving displaced peoples' living conditions is only in short-term and limited. Recently, several INGOs have been involved in facilitating the government of Vietnam to formulate a benefit sharing policy.

For example, Winrock International is an US INGO with an office in Ha Noi and is officially registered with VUFO to operate in Vietnam. Since 2006, Winrock International has been recognized as a pioneer in promoting and facilitating the benefit sharing mechanism through payment for forest environmental services (PFES) in collaboration with Vietnamese government, local authorities, international donors, and local communities. In this mechanism, communities displaced due to hydropower dam construction have recently benefited from the PFES programs. Between 2006 and 2007, the Asia Regional Biodiversity Conservation Program (ARBCP) of Winrock International supported the province of Lam Dong to develop its first Biodiversity Conservation Action Plan (BCAP), which was the first plan in Vietnam fully incorporated into a province's socio-economic development plan. This plan initiated a fund to collected revenues from payments from forest environmental services to sustain the province's priority on biodiversity conservation targets (Winrock International, 2011, p. 7). From this success, in 2007, the Ministry of Agriculture and Rural Development (MARD) invited Winrock to help develop a pilot policy for Payment for Forest Environmental Services in Vietnam. Winrock carried out baseline assessment in Da Nhim hydropower watershed to identify the economic value of ecosystems services paid to each forest environmental unit and for each type of environmental service buyer. This collaboration between MARD and Winrock led to the first policy on PFES in Vietnam, approved under Prime Minister's Decision 380/QĐ-TTg (April 2008), with activities undertaken through December 2010. MARD was the leading ministry to implement this pilot policy in two provinces: Lam Dong in the south and Son La in the north, where Winrock was selected to undertake implementation in Lam Dong Province, while the German development agency (GTZ) supported activities in Son La Province from 2009 to 2010 (Winrock International, 2011, p. 12). As the focus of this paper is on INGO involvement, Winrock International involvement was our focal analysis.

In accordance to decision 380/QĐ-TTg, Winrock established a network of partners from central to local-level governments and provided financial and technical supports to responsible ministries, provinces and their functional departments, and local authorities, as well as beneficiaries to complete the pilot from preparation, implementation, monitoring, and evaluation phases. In Lam Dong province, Winrock worked with the Lam Dong Provincial Peoples' Committee (LDPPC), Department of Agriculture and Rural Development (DARD), and MARD's Forest Protection Department (FPD) to design and develop a PFES program (See Figure 6-2). A national PFES Steering Committee was formulated with representatives from MARD, Lam Dong DARD, Lam Dong PPC, Office of the Government, Ministry of Planning and Investment (MPI), Ministry of Finance (MOF), and Winrock. The committee's primary purpose is to evaluate, direct, and monitor the implementation of PFES pilots. With support from USAID/RDMA (Regional Development Mission for Asia), Winrock played a key role in designing and implementing the pilot in Lam Dong province.

From this network, Winrock obtained a strong consensus and commitment from central government to local authorities to implement the pilot. Under this support, it was effective for Winrock to contact with the payers and forest environmental service providers (beneficiaries) to get an agreement of these stakeholders. During preparation phase between 2007 and 2008, Winrock conducted a variety of research and consultation with experts in the region and internationally, with the aim of informing policy development processes, providing a scientific basis for the social and economic basis for the pilot selection, and determining a market value for water, soil, and aesthetics environmental services in the Dong Nai river basin. The results of these studies set the payment levels at 20 VND per kilowatt-hour from commercial hydropower production businesses, 40 VND per cubic meter from clean-water production businesses, and 0.5%-2% of annual revenue from tourism businesses (Winrock International, 2011). However, to build consensus with stakeholders, Winrock conducted activities to raise awareness about PFES for officials (from central to local), payment companies, and for forest environment service providers because PFES is a relative new concept in South East countries and Vietnam in particular. In 2007, Winrock worked with the US Forest Service (USFS) to support representatives of Vietnam, Thailand, and Cambodia in a study in US to learn successful models of PFES. In 2008, Winrock also collaborated with district and commune authorities to carry out community-level public awareness enhancement in the pilot region. Furthermore, Winrock worked with local news agencies to broadcast these events and other PFES stories in their weekend programs and in local (ethnic-minority) languages to ensure a wide audience among local farming communities. PFES public information campaigns provided 24 poster panels, 200 small billboards at major tourist sites, 41 large roadside billboards, and 14,200 brochures. Winrock also assisted MARD to produce a 30-minute video on PFES implementation to propagate on national television and at various inter-ministerial and national meetings. To carry out the tasks of the pilot, Winrock implemented a series of capacity-building activities for project staff and ten members of the Lam Dong PFES Technical Working Group. Winrock organized a series of workshops and discussion groups with international PES experts. Provincial technicians were also supported in attending overseas courses. Also, Winrock facilitated over 50 technical training courses for officials and technicians from 15 other provincial agencies (Winrock International, 2011).

In 2009, the final PFES pilot was finalized. Service payers were identified under the decision of national PFES Steering Committee with the assistance of Winrock, including two hydropower companies (Da Nhim and Dai Ninh hydropower plants), Sai Gon water company (SAWACO), Dong Nai water company (DNWACO), and tourism operators in Lam Dong province, of which the two hydropower companies were selected to carry out the pilot. The payment value of Da Nhim and Dai Ninh for PFES pilots was about US$2.3 million (48 billion VND) according to the estimate of Winrock. This money was deposited to the Lam Dong Forest Protection and Development Fund (FPDF), which was established by Lam Dong province with the technical assistance of Winrock. This was the first benefit-sharing fund from FPES in Vietnam.

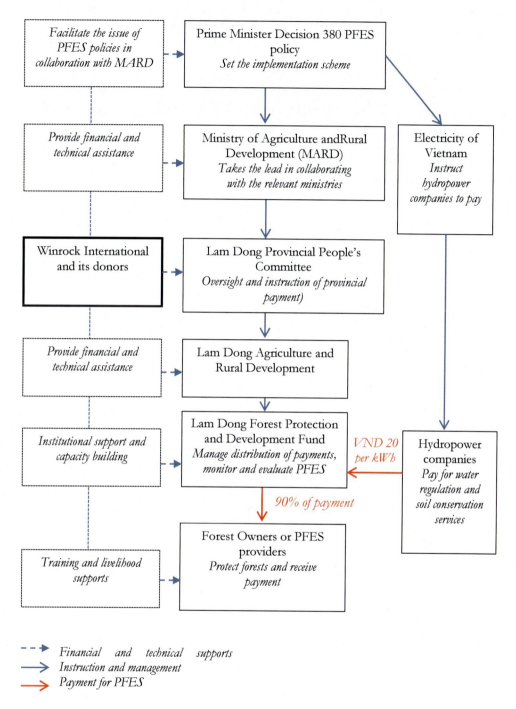

Figure 6-2 Institutional network to implement the policy pilot of PFES in Lam Dong province, Vietnam. *Source: Adapted from Winrock, 2011*

The fund is supervised by an independent governing board consisting of representatives from DARD, Department of Finance, DONRE, Department of Planning and Investment, Department of Taxation, Department of Industry and Trade, and the State Treasury, who, together, appoint the fund's manager. The fund is monitored by independent auditors to ensure transparent and proper use of the monies received, as well as proper pay outs to the forest protection services. This payment mechanism was also formulated to collect money from PFES payers and providers (See Figure 6-2). After establishing FPDF, Winrock facilitated all stakeholders to sign the Memorandum of Understandings (MoUs), including Vietnam Electricity Group (EVN), SAWACO, DNWACO, and 59 tourism companies who committed to pay US$3.4 million corresponding to 220,000 hectares of forests protected by PFES providers or PFES beneficiaries (Winrock International, 2011).

According the evaluation of Winrock and MARD in 2010, PFES pilot policy was being implemented quite successfully, with the income of rural households increasing slightly. The first pilot contracted was with 218 households in five communes for patrolling 4,795 hectares of forest in the Da Nhim watershed. In 2009, each hectare of protection forest was paid US$14 (290,000VND) per year, but increased to US$19 (400,000VND) per year in 2010. Subsequently, Winrock asked FPDF to expand the pilot to 22 Forest Management Boards and forest enterprises, as well as to 9,870 households, in which 6,858 households were ethnic minorities. They contracted to protect 220,000 hectares of threatened forests. Each household received from US$438 to US$470 per year. Each household contracted to protect approximately 30 ha. In 2010, payment rate increased and, therefore, each household received higher payment, from US$540 to US$615 annually. The payment rate of FPES programs was 400% higher than payment for forest protection implemented by the Vietnamese government previously implementation. PFES became an important source of income for poor households, especially those of ethnic minorities (Winrock International, 2011).

Additionally, forests in the pilot areas were acknowledged to be better protected, with the incidence of violations and encroachment reduced significantly. Winrock provided technical and financial supports to Lam Dong Provincial Peoples' Committee (PPC) to establish a monitoring system to monitor the effectiveness of PFES on biodiversity and environment. This system involved most communes, wards, and districts in the pilot areas. Information in the patrolling logbooks showed that forest protection services of local people reduced by 50% illegal logging and wildlife poaching in the Da Nhim watershed (Winrock International, 2011).

The success of PFES pilots in Lam Dong province was the key catalyst for government to confidently embark on more pilots in 15 further provinces and to develop a comprehensive legal framework for PFES in Vietnam. In 2010, the Prime

Minister issued Decree 99/2010/CP-TTg to regulate the implementation of PFES. This is the first legal framework to share benefits from economic activities to households who are protecting forests, in which resettled households are also involved in FPES models. In 2011, MARD also issued Circular 80/2011/TT-BNNPTNT (November 2011) to provide guidelines for determining payment rates for forest environmental services with the technical support of Winrock. It was the first time that an INGO intervened deeply in making a national policy successfully in Vietnam, where Winrock made significant influence on the government in Vietnam. However, Winrock also faced a number of barriers in partnership with state agencies in the process of implementing pilot. The roles of MARD and MONRE were not clear, which constrained collaboration. Also, coordination from central to local authorities was still not effective. Subordinate technical agencies of provinces and ministries did not often work closely and in a timely manner, resulting in delays of implementation, especially the allocation of budget from central to provincial and district levels. This barrier is similarly recognized as operational constraint for the partnership between GOs and INGOs in the study of Nguyen (2011).

Pilot extension of payment for forest environmental services in Quang Nam province: The success of these pilots in Lam Dong province helped the government and Winrock embark on expanding the FPES model in Quang Nam province, Central Vietnam. Similar to the pilot in Lam Dong, Winrock carried out technical and financial assistance to develop PFES programs. Two non-resettled villages were selected to implement FPES in 2010. The institutional arrangement and payment mechanism from Lam Dong province was applied. Until 2013, two resettled villages were selected to participate in FPES models, including A Den and Tro Gung resettled villages displaced by A Vuong hydropower dam in 2005. Each resettled household received 22 hectares of forest to protect using FPES model. One person from each household is expected to provide approximately one day's labour per week. A team includes a group of households established to patrol illegal loggers in collaboration local forest rangers of the A Vuong forest protection management unit. According to the village head and district officials, the number of illegal forest loggings decreased by 50% compared to before FPES program. In Aden village in Dong Giang district, resettled households were divided into four teams with 13 households in each group for monitoring. Each household received US$13 (274,000VND) per ha a year, so each household received an annual income of US$256 (6 million VND) or US$24 (502,333 VND) per month. This income enabled many resettled household to escape from poverty. However, the payment rate in Quang Nam was less than the rate of the pilots in Lam Dong province (Winrock International, 2011; Singer et al, 2014).

According to an evaluation by local officials, the PFES model in Quang Nam province was not as successful as the pilot in Lam Dong province. First, the leader of district forest protection management unit (DFPMU) said that transaction costs were high, including costs for implementation of an initial baseline study to determine initial forest cover, quality and composition, and periodic forest monitoring. Tax revenues paid by hydropower plants were often insufficient to pay households who protected forests. Additionally, hydropower companies often delayed payment to the Provincial Forest Protection and Development Fund; as a result it was challenging to pay displaced people involved in the PFES programs (In-depth interview with DFPMU, 2012). According the Forest Protection and Development Fund of Quang Nam province, hydropower and water supply companies must pay to the fund about US$524 million (11 billion VND), but most did not pay to the fund. For example, A Vuong hydropower has failed to pay US$352 million (7.4 billion VND) (Tran Huu, 2014). For resettled households, the major restriction for participation is the requirement that they forego traditional swidden practices of converting protected forest land to upland rice or other types of agriculture to secure needed food security. However, in this case, forest land that had already been converted to shifting cultivation was excluded from demarcated forest areas to allow for continued use (Singer et al., 2014). Additionally, service providers of PFES schemes had no opportunity to negotiate with FPES payers regarding the payment rate because Decree 99/2010/ND-CP applied a K-coefficient for each forest type and forest cover to determine payment rates. Therefore, it was not fair for households who received poor forests because they protect the same forest land area, but received less income per hectare than other households (Catacutan et al., 2011).

Also, there are concerns for both the environmental and economic sustainability of the PFES scheme in Dong Giang district. With most residents eager to engage in rent-seeking behaviour on their allocated land by planting fast-growing acacia trees, there may be adverse impacts on local biodiversity with monoculture plantations. Sporadic or limited inspection of forest conditions by program managers may reduce motivation for FPES providers to thoroughly monitor the condition of remote or inaccessible parts of their forest allotment. PFES schemes have been regarded by some local officials and researchers as a "silver bullet" policy instrument for dam-affected communities (McElwee, 2012). However, the PFES initiative was regarded as a modest welfare program that secures a small income stream for FPES providers but not substantial economic security, providing insufficient training and support to enable FPES to gain new livelihood skills or enhance future incomes. The program's complexity, high costs, and the necessity of untangling competing tenure claims to forest land may also pose challenges to implementation for all resettled communities (To, et al., 2012).

Thus, it can be seen that the relationship between INGOs and the state was established on the basis of shared objective and a mission toward poverty reduction and environmental protection, as well as sustainable development. Sharing of common goals created a strategic partnership between INGOs and the party-state. The active and constructive participation of INGOs in Vietnam's development policies helped INGOs increase their influence level toward policy development of Vietnam, especially to resettled communities who have benefited from FPES schemes. The role of INGOs has been transferred to the new page because, today, they can intervene deeply into formulating and implementing new development policies based on their great financial and technical capacity that the government of Vietnam is not able to perform well.

6.5   Discussion and conclusion

The findings of this study reveal the positive relationship of VNGOs and INGOs with the party-state of Vietnam and local authorities. The door has been gradually opening for NGOs to influence the decision-making process of government on hydropower development. Government has tolerated, endorsed, and recognized important roles of NGOs in the governance framework of hydropower development. This context changed the political views of many people (Nga Dao, 2010; Ha, 2012; CODE, 2010) who asserted that hydropower development and land acquisition and resettlement are politically sensitive issues and difficult for NGOs to intervene in Vietnam. In practice, NGOs have brought both positive partnerships with and changed political views of the state on hydropower development, with appropriate intervention approaches. For example, they applied evidence-based lobbying and advocacy, followed the rules of law, and handled the needs and concerns of various interest groups to government.

VNGOs enhanced the responsiveness and accountability of the state on addressing adverse consequences to environment and society and gradually improved the degree of transparency of the decision-making process of building hydropower dams. At the local level, VNGOs also strengthened local authorities' responsiveness and accountability through improving the efficiency of social services, acting as a strategic partner for reform-oriented collaborators and monitors, and promoting grassroots democracy through the mobilization of local people to be involved in asking local authorities to provide sufficient land and public services after resettlement. More importantly, VNGOs strengthened civil society through anti-dam movement. They created a clearer space or arena for civil society organizations to negotiate and debate with government and other stakeholders to insist that government incorporate good governance into hydropower development policy. Additionally, VNGOs pressured the state to endorse their legitimacy in Vietnamese society by requesting the government formulate legal framework, so that VNGOs are protected by legitimate laws for their actions and campaigns. As a result, the government endorsed the appearance of VNGOs in Vietnamese society as official and legal organizations. This is the reason for VNGOs ability to sustain their missions and actions on the ground, which is very different from actions of other civil society organizations led by elite groups that are often dismiss after contested issues were resolved (Bui, 2013).

VNGOs are working with staffed employees and in close partnership with government, local authorities, private sectors, INGOs, and local people to strengthen their power in influencing the course of development in Vietnam. The literature review shows that principles and actions of NGOs often expressed as threats to state's power and legitimacy in the provision of social services (Fowler, 1991; Tvedt, 1998; Marcussen, 1996). The case of hydropower development in Vietnam reflect that the principles and actions of NGOs do not threaten or weaken the role and power of the state; in contrast, they increase the legitimacy and power of the party-state because their oppositional actions are also a good opportunity for the party-state to express their political will in reforming the institutional as committed to the entire people. The victory of VNGOs is also a triumph of the party-state in front of the entire people. The party-state took this opportunity to strengthen their legitimacy on "the rule of law" and constitutionalism as indicated by Thayer (2009, p. 48).

International non-government organizations (INGOs) also play a very important role in hydropower development in Vietnam. As a strategic partner with the state for a long time, INGOs developed new initiatives to distribute wealth to vulnerable people and to protect environment. The intervention of Winrock International is an example. They initiated the first benefit-sharing framework in Vietnam though the development of the forest environment payment service model. Their successful pilots expanded not only throughout Vietnam but spread out to South East Asian countries, such as Laos, Cambodia, and Thailand. Based on their success, the government of Vietnam appreciated and agreed to formulate the first benefit sharing mechanism for displaced people due to hydropower dam development. It shows that INGOs have made a big influence on the development policy of Vietnam.

Thus, we can see that NGOs have played important roles and made significant influence on the development of hydropower in Vietnam. They become opposite actors against government and local authorities, but they are also constructive partners with government and local authorities because they share the same vision about sustainable development. However, the question ahead: are there barriers between the state and NGOs? Are NGOs totally free, autonomous, and independent to operate in any development policies of Vietnam? Are they able to overcome? Vietnam is as a single party-state that with power located in leaders of government and local authorities. The government often claims to provide all social services to the people through government organizations and its mobilization associations by applying the monopoly mechanism (Heurlin, 2010; Bui, 2013; Kelkvliet, 2001; Landau, 2010). Therefore, many development areas are monopolized by the state and its corporations. As a result, development decisions are still made between state agencies and state-owned companies where the peoples' participation is excluded. Many large-scale development projects, such as land acquisition for foreign land lease, golf courses, resorts, industrial parks, and infrastructure development, have caused adverse consequences on environment and society similar to hydropower development. However, there are a limited number of VNGOs and INGOs who influence these decisions of the government and local authorities. With this in mind, the success of VNGOs and INGOs in hydropower development to open up a new era for civil society in Vietnam, in which VNGOs and INGOs should participate in all development activities of the party-state, government, and local authorities, is an example of a positive movement forward for society as a whole. In doing so, the objective of sustainable and equitable development may be achieved.

References

Baker, G. (1999). The taming of the idea of civil society. *Democratization, 6*(3), 1-29.

Biswas, A. K. (2012). Impacts of Large Dams: Issues, Opportunities and Constraints. In *Impacts of Large Dams: A Global Assessment* (pp. 1-18). Springer Berlin Heidelberg.

Bratton, M. (1994). *Civil society and political transition in Africa* (pp. 51-81). Institute for Development Research.

Bui, T. H. (2013). The development of civil society and dynamics of governance in Vietnam's one party rule. *Global Change, Peace & Security, 25*(1), 77-93.

Bui, T. H. (2013). The development of civil society and dynamics of governance in Vietnam's one party rule. *Global Change, Peace & Security, 25*(1), 77-93.

Catacutan, D. C., Ha, H. M., Sen, H., & Luan, T. D. (2011). Moving beyond pilots: a review of lessons learnt in payments for forest ecosystem services (PFES) in Vietnam. *World Agroforestry Center.*

Charnovitz, S. (1996). Two centuries of participation: NGOs and international governance. *Mich. J. Int'l L., 18*, 183.

CIEM and JICA. (2003). Study on Donor Practices in Vietnam: Grant Aid and Transaction Costs -Listen to the Voice of the Recipient. Retrieved from http://www.jica.go.jp/cdstudy/ library/pdf/C20071101 01.pdf

Clarke, G. (1998). Non-Governmental Organizations (NGOs) and Politics in the Developing World. *Political studies, 46*(1), 36-52.

CODE. (2010). *Displacement, resettlement, living stability and environmental and resources protection in hydropower dam projects.* CODE: Ha Noi, Vietnam.

Cohen, J. L. (1994). *Civil society and political theory.* Mit Press.

Cohen, J., & Arato, A. (1992). Politics and the Reconstruction of the Concept of Civil Society. *Cultural-political interventions in the unfinished project of enlightenment,* 120-142.

Cuong, B. T. (2005). Civic Organizations in Vietnam. Institute of Socialogy: Ha Noi, Vietnam

Dalton, R. J., & Ong, N. N. T. (2005). Civil society and social capital in Vietnam. *Modernization and Social Change in Vietnam. Hamburg, Institut für Asienkunde.*

Dao, N. (2010). Dam development in Vietnam: the evolution of dam-induced resettlement policy. *Water Alternatives, 3*(2), 324-340.

Dao, N. T. V. (2012). *Resettlement, Displacement and Agrarian Change in Northern Uplands of Vietnam.* York University.

Diamond, L. J. (1994). Toward democratic consolidation. *Journal of democracy, 5*(3), 4-17.

Diamond, L. J. (1994). Toward democratic consolidation. *Journal of democracy,5*(3), 4-17.

Dong, N. (2013). Escape from hydropower projects. Retrieved from http://nld.com.vn/kinh-te/thao-chay-khoi-cac-du-an-thuy-dien-20130829095220379.htm

Ekiert, G., & Hanson, S. E. (Eds.). (2003). *Capitalism and democracy in Central and Eastern Europe: Assessing the legacy of communist rule.* Cambridge University Press

Falk, R. (1995). World Order between Interstate Law and the Law of Humanity: The Role of Civil Society Institutions, The. *Int'l Legal Theory, 1*, 14.

Fisher, J. (1998). *Nongovernments: NGOs and the political development of the third world* (pp. 11-12). West Hartford, CT: Kumarian Press.

Foster, K. W. (2001). Associations in the embrace of an authoritarian state: state domination of society?. *Studies in Comparative International Development, 35*(4), 84-109.

Foster, K. W. (2002). Embedded within state agencies: Business associations in Yantai. *The China Journal,* 41-65.

Fowler, A. (1991). The role of NGOs in changing state-society relations: perspectives from Eastern and Southern Africa. *Development Policy Review,9*(1), 53-84.

Geddes, A. (1999). Queen's Papers on Europeanisation No 4/2003 Still Beyond Fortress Europe? Patterns and Pathways in EU Migration Policy. *Official Journal No. C, 19*, 1.

Goodman, R., White, G., & Kwon, H. J. (Eds.). (1998). *The East Asian welfare model: Welfare orientalism and the state.* Psychology Press.

Gray, M. L. (1999). Creating civil society? The emergence of NGOs in Vietnam. *Development and Change, 30*(4), 693-713.

Ha, N. (2010, 06 11). *Forest land lease to foreigners: I do not accept this idea.* Retrieved from http://vneconomy.vn/thoi-su/cho-nuoc-ngoai-thue-dat-trong-rung-quan-diem-cua-toi-la-khong-20100611025144480.htm

Ha, T.V. (2011). Local People's Participation in Involuntary Resettlement in Vietnam: A Case Study of the Son La Hydropower Project. In K.LAZARUS, B. NATHAN, D. NGA & B.P. RESURRECCION, *Water rights and social justice in the Mekong Region.* Routledge.

Habermas, J. (1991). *The structural transformation of the public sphere: An inquiry into a category of bourgeois society.* MIT press.

Hakkarainen, M., & Katsui, H. (2009). Partnership' between Northern NGOs and Vietnamese Counterpart Organisations. In T. Veintie, & P. Virtanen, *Local and Global Encounters: Norms, Identities and Representations in Formation* (pp. 117-136). Helsinki: Renvall Institute Publiations 25.

Hang, B. T. (2006). *The Revised Civil Code and its Impacts on NGOs.* VNAH-SPF Thematic Research 6.

Hansson, E. (2003). Authoritarian Governance and Labour: The VGCL and the Party-State in Economic Renovation. *Ben J. Tria Kerkvliet, Russell HK Heng and David WH Koh, eds., Getting Organized in Vietnam: Moving in and around the Socialist State. Singapore: Institute of Southeast Asian Studies,* 153-184.

Heurlin, C. (2010). Governing civil society: The political logic of NGO–state relations under dictatorship. *Voluntas: International Journal of Voluntary and Nonprofit Organizations, 21*(2), 220-239.

Huu, T. (2014). *Forest environmental service payment: need a strict regulation for late payment.* Retrieved from http://baoquangnam.com.vn/kinh-te/lam-nghiep/201405/chi-tra-dich-vu-moi-truong-rung-can-co-che-tai-xu-ly-manh-482033/

International River Network. (2014). *Southeast Asia Partner Organizations.* Retrieved from http://www.internationalrivers.org/resources/southeast-asia-partner-organizations-3598

Kerkvliet, B. (1994). Politics of Society in Vietnam in the mid-1990s. In *ANU Vietnam Update Conference, Canberra* (pp. 396-18).

Kerkvliet, B. J. T. (2001). An approach for analysing state-society relations in Vietnam. *Sojourn: Journal of Social Issues in Southeast Asia,* 238-278.

Kiên, B., & Tuyên, P. (2010). *Forest land leased to foreign countries is very dangerous.* Retrieved from Forest land leased to foreign countries is very dangerous

Landau, I. (2008). Law and civil society in Cambodia and Vietnam: A gramscian perspective. *Journal of Contemporary Asia*, *38*(2), 244-258.

Lehmbruch, G., & Schmitter, P. C. (1982). *Patterns of corporatist policy-making*. Sage.

London, J. (2004). Rethinking Vietnam's mass education and health systems. *Rethinking Vietnam*, 127-142.

Ma, Q. (2002). Defining Chinese nongovernmental organizations. *Voluntas: International Journal of Voluntary and Nonprofit Organizations*, *13*(2), 113-130.

Ma, Q. (2005). *Non-governmental organizations in contemporary China: paving the way to civil society?*. Routledge.

Marcussen, H. S. (1996). NGOs, the state and civil society. *Review of African Political Economy*, *23*(69), 405-423.

Marr, D. (1994, November). The Vietnam Communist Party and Civil Society. In *Vietnam Update 1994 Conference: Doi Moi, the state and civil society* (pp. 10-11).

McCormick, J. (1999). The role of environmental NGOs in international regimes. *The global environment: Institutions, law, and policy*, 52-71.

McElwee, P. D. (2012). Payments for environmental services as neoliberal market-based forest conservation in Vietnam: Panacea or problem?.*Geoforum*, *43*(3), 412-426.

Meidinger, E. (2001, June). Law Making by Global Civil Society: the Forest Certification Prototype. In *Paper Presented for the Conference on Social and Political Dimensions of Forest Certification*.

Mercer, C. (2002). NGOs, civil society and democratization: a critical review of the literature. *Progress in development studies*, *2*(1), 5-22.

Metzger, T. A. (1998). *The Western concept of a civil society in the context of Chinese history*. Hoover Press.

MOIT. (2013). *Kết quả rà soát quy hoạch, đầu tư xây dựng và vận hành các dự án thủy điện trên cả nước*.

Nguyen, A. N. T. (2011). *A Case Study of NGO-Government Collaboration in Vietnam: Partnership Dynamics Explained through Contexts, Incentives, and Barriers* (Doctoral dissertation, Texas A&M University).

Le, T. N. (2011). Role and challenges from irrigation and hydropower dams in Central Vietnam. *Journal of Science*, *68*, *79-88*.

Nørlund, I. (2007). *Filling the Gap: The Emerging Civil Society in Viet Nam*. Viet Nam Union of Science and Technology Associations.

Oxhorn, P. (1995). *Organizing civil society: The popular sectors and the struggle for democracy in Chile*. University Park: Pennsylvania State University Press.

Pekkanen, R. (2006). *Japan's dual civil society: Members without advocates*. Stanford University Press.

Phuoc, D. (2014, May 5). *National conference on the prevention and fight against corruption.* Retrieved from http://noichinh.vn/cong-tac-phong-chong-tham-nhung/201405/hoi-nghi-toan-quoc-ve-cong-tac-phong-chong-tham-nhung-294534/

Pitkin, H. F. (1967). *The concept of representation.* Univ of California Press.

Ramasamy, P. (2004). Civil Society in Malaysia. *Civil Society in Southeast Asia,* 198.

Riker, J. V. (1995). Contending perspectives for interpreting government-NGO relations in South and Southeast Asia: Constraints, challenges and the search for common ground in rural development. *1995), Government-NGO Relations in Asia, MacMillan Press Ltd, London,* 15-55.

Sakai, Y. (2002). Flexible NGOs vs Inconsistent State Control. *The State & NGOs: Perspective from Asia,* (25), 161.

Sakata, S. (2006). Changing Roles of Mass Organizations in Poverty Reduction in Vietnam. *Actors for poverty reduction in Vietnam,* 49-79.

Salamon, L. M., & Anheier, H. K. (1998). Social origins of civil society: Explaining the nonprofit sector cross-nationally. *Voluntas: International Journal of Voluntary and Nonprofit Organizations, 9*(3), 213-248.

Schmitter, P. C. (1974). Still the century of corporatism?. *The Review of politics, 36*(01), 85-131.

Singer, J., Ty, P. H., & Hai, H. (2014). Broadening stakeholder participation to improve outcomes for dam-forced resettlement in Vietnam. *Water Resources and Rural Development.*

Sinh, B. T. (2002). Government and NGO partnership in managing community-based water resources in Vietnam: a case study of Thai Long Dam Project. *Business Strategy and the Environment, 11*(2), 119-129.

Sinh, B. T. (2003, February). Civil society and NGOs in Vietnam: some initial thoughts on developments and obstacles. In *Paper presented at the Meeting with the Delegation of the Swedish Parliamentary Commission on Swedish Policy for Global Development to Vietnam* (Vol. 3, p. 2002).

Su, N., & Thuy, H. (2014, May 5). *General Secretary of Vietnam Communist Party: The civil organizations-based anti-corruption.* Retrieved from http://www.vietnamplus.vn/tong-bi-thu-chi-dua-vao-dan-moi-chong-duoc-tham-nhung /258055.vnp

Thang, N. T. (2010, November 10). The policy of socialization on the environment and preferential policies for investment in the environmental sector of Vietnam today. Retrieved from http://vea.gov.vn/vn/truyenthong/sukien-ngayle/hoinghimttq /xuctiendautu/Pages/

Thayer, C. A. (1992). Political reform in Vietnam: Doi moi and the emergence of civil society. *The development of civil society in communist systems,* 110-129.

Thayer, C. A. (1995). Mono-organizational socialism and the state. *Vietnam's rural transformation*, 39-64.

Thayer, C. A. (2009). Vietnam and the challenge of political civil society. *Contemporary Southeast Asia: A Journal of International and Strategic Affairs,31*(1), 1-27.

Thayer, C. A. (2010). Political legitimacy of Vietnam's one party-state: Challenges and responses. *Journal of Current Southeast Asian Affairs, 28*(4), 47-70.

To, P. X., Dressler, W. H., Mahanty, S., Pham, T. T., & Zingerli, C. (2012). The prospects for payment for ecosystem services (PES) in Vietnam: a look at three payment schemes. *Human Ecology, 40*(2), 237-249.

Tvedt, T. (1998). *Angels of mercy or development diplomats? NGOs & Foreign Aid*. James Currey Ltd. & Africa World Press, Inc..

Ty, P. H., Phuc, N. Q., & Westen, G. V. (2014). Vietnam in the debate on land grabbing: conversion of agricultural land for urban expansion and hydropower development. In M. Kaag, & A. Zoomers, *The Global Land Grab: Beyond the Hype* (pp. 135-151). London: ZED Book.

Ty, P. H., Van Westen, A. C. M., & Zoomers, A. (2013). Compensation and Resettlement Policies after Compulsory Land Acquisition for Hydropower Development in Vietnam: Policy and Practice. *Land, 2*(4), 678-704.

UNDP in Vietnam. (2006). The Emerging Civil Society An Initial Assessment OF Civil Society in Vietnam. *Structure, 2, 3.*

Vietnam administration of forestry. (2014). *Afforestation scheme for forest losses for other purposes*. Retrieved from http://tongcuclamnghiep.gov.vn/tin-tong-cuc/nam-2016-hoan-thanh-trong-rung-thay-the-khi-chuyen-muc-dich-su-dung-rung-sang-muc-dich-khac-a1694

Vietnam Government. (2014, January 01). *New Year's message, Prime Minister Nguyen Tan Dung*. Retrieved, from http://baodientu.chinhphu.vn/Cac-bai-phat-bieu-cua-Thu-tuong/Thong-diep-nam-moi-cua-Thu-tuong-Nguyen-Tan-Dung/189949.vgp

Vietnam River Network (VRN). (2013). *Report of Environmental Impact Assessment on Dong Nai 6 and 6A hydropower dams*. Vietnam: VRN

Warren, M. E. (2001). *Democracy and association*. Princeton University Press.

Winrock International. (2011). *Payment for Forest Environmental Services: A Case Study on Pilot Implementation in Lam Dong Province Vietnam from 2006 - 2010*. Vietnam: Winrock International.

Wischermann, J., & Nguyen, Q. V. (2003). The relationship between civic and governmental organizations in Vietnam: Selected findings. *Getting organised in Vietnam: Moving in and around the socialist state*, 185-233.

7 CONCLUSION: TOWARD BETTER GOVERNANCE FOR EQUITABLE AND SUSTAINABLE HYDROPOWER DEVELOPMENT

Hydropower, as a renewable and sustainable energy source, has been promoted as an important catalyst to promote the high ambition of government for industrialization and modernization for many developed countries before 1975. Since mid-1970s, many developing countries have aggressively advanced the development of hydropower for the sake of electricity security and economic growth. However, the value of hydropower has been rethought over time because of its widely negative impacts on environment and society that constraints equitable and sustainable development to date.

This research looked at the case of hydropower development in Vietnam as an example to explain why it has emerged as a developmental issue and to search for solutions for governing equitable and sustainable development. After independence in 1945, Vietnam, a very poor country, started growing with a severe electricity shortage. However, Vietnam has employed the advantage of an abundant water resource potential to construct hydropower dams to produce enough electricity for socio-economic development. However, as I have argued, many people receive the benefits of hydropower dam construction and promote further development, but others assert that Vietnam should discontinue hydroelectricity dam construction because they are very harmful for society and environment. Therefore, I investigated a number of aspects of hydropower development in Vietnam, including benefits and costs of hydropower development and governance issues of land acquisition, compensation, displacement and resettlement, and benefit-sharing policy. Each topic has been clarified in each chapter, and many have been incorporated into different chapters, particularly nine elements of governance, including participation, rule of law, transparency, responsiveness, consensus orientation, equity, effectiveness and efficiency, accountability, and strategic vision have been spread over all chapters of the dissertation. Based on previous chapters, the conclusion aims to answer the given central question: *Under what conditions can hydropower dam development contribute to equitable and sustainable development?*

The answer for this question is highlighted in four arguments. First of all, I argue that the outcomes of hydropower development in Vietnam have shown four dilemmas of sustainable development. Secondly, the political system of Vietnam has created a large power gap between the power holders (government, local authorities, and hydropower developers) and the powerless (displaced people and other indirectly affected people due to dam construction). The power imbalance in hydropower development making-decisions obstructed the implementation of good governance on the ground. Thirdly, I will argue that the strong resistance and advocacy of Vietnamese NGOs has increased the tolerance and endorsement of the party-state for civil society organizations to involve in decisions-makingon hydropower. This has shown positive signs for improved governance on the ground. Finally, I argue that less-developed countries need to rethink their hydropower development plans, and it is necessary for international organizations to re-adjust their strategies to introduce good practices of hydropower dam construction and good governance models to less-developed countries to meet the need of each country acceptably and effectively and in a timely manner. First of all, I will reflect four dilemmas of hydropower development in Vietnam.

7.1   Dilemmas of sustainable hydropower development in Vietnam

The development perspective of the Vietnamese government is that economic growth must tie-in with environmental protection and social equity. However, the case of hydropower development suggests that hydropower dam construction has shown four developmental paradoxes obstructing the sustainability of development in Vietnam, as follows.

Firstly, satisfaction of indispensable needs of electricity versus conservation of natural resources. The research findings of Chapter 2 reflect that Vietnam has been successful in the exploitation of hydropower potential to meet increasing demand of electricity for industrialization and modernization. So far, hydroelectricity and other power sources have provided sufficient electricity for the whole country, from urban to rural areas. However, Vietnam has had to pay a high price for the loss of valuable ecosystems submerged in hydropower reservoirs, such as large areas of fertile lands along rivers, which are crucial for food production, rich forests, and rare faunas and floras. Many hydropower companies also destroy natural forests for constructing transmission lines, offices, roads, and other structures. Furthermore, natural forests are increasingly destroyed by resettled people, who clear forests for slash-and-burn cultivation because they do not have enough land in new resettlement sites. Meanwhile, hydropower companies and local authorities do not replant to offset lost forest areas. Thus, Vietnam satisfied electricity needs for economic growth but lost forests and valuable ecosystems.

Secondly, increased production of wealth to support growing needs versus fair distribution of accumulated wealth. Chapter 2 also shows that hydroelectricity has contributed significantly to the economic growth of the country. Its contribution has increased significantly the value of industry and service sectors to

162

meet the target of industrialization and modernization of Vietnam to2020. As a result, Vietnam has become a middle-income country. Electricity security has contributed to improving people's living conditions from rural to urban areas across the country, and, along with this, the quality of technical and social infrastructure. Hydropower also has brought considerable income to the state, local governments, hydropower companies, and electricity service companies and their staffs. However, that wealth has not been fairly redistributed to those displaced to make way for hydropower dam construction. Particularly, most displaced people are ethnic minority groups living in mountainous regions with low living conditions that are very vulnerable to the risks of displacement and resettlement. They sacrificed many things, including land, houses, health, culture, and living spaces where are embedded many important spiritual values to them (See chapters 4, 5).

But, what benefits did displaced people get from the development of hydropower dams? The findings of chapter 5 revealed that displaced people had little benefit from the development of hydropower. Although they received compensation and assistances, these not adequate compared to their losses of land, homes, income, job, and common pool resources. The supports for livelihood restoration and development were short-term and not effective enough to enable to them have at least equal living conditions as before displacement. More often, most displaced communities resettled in places where land and natural resources were depleted. More importantly, there was no benefit sharing mechanism to redistribute the revenues of hydroelectricity to support resettled communities to improve their livelihoods and living conditions. They received only one or two years of electricity subsidy, but after that they had to pay for electricity (See chapters 2, 3, 4, and 5). This was an ironic situation when displaced people sacrificed so much as compared the benefits they received from hydropower development. As a result, they could not restore their traditional livelihoods and had few opportunities to develop new livelihood strategies and their living conditions often became worse than before displacement; they were often poorer than non-displaced communities in the same region (See chapters 4, 5).

This situation reflects that the growing wealth of whole country does not guarantee its fair distribution to displaced persons who make way for growth. The economic growth of country has reduced significantly the national poverty rate as a whole, but intensified the poverty rate of displaced communities (chapters 2, 3, 4, and 5). As promised by the Vietnamese government, the objective of resettlement is to ensure displaced people have at least an equal or better life after displacement, but the results show the failure of implementation on the ground. Thus, the trade-off between the hydropower-based economic growth and social protection (equity for displaced people) is not operated well in practice to ensure equitable development.

***Thirdly, inequity between groups of people: Rights of small numbers of locally
affected populations versus rights of larger numbers of potential beneficiaries.***
In Vietnam, hydropower dam construction is considered for the sake of public
development. By law, water, land, and other natural resources used for hydropower
dam construction are owned by the entire people of the country. The government
represents all people in the management of all natural resources and is responsible for
the development of natural resources for the entire people. However, when making
decisions on hydropower development in practice, the government often looks
towards benefits for the majority of people and not to marginalized and displaced
communities who account for only 0.3% of the total population. Additionally, many
hydropower dams violate the rights of people living in downstream regions who
depend on water resource from upstream for agriculture, fishing, water supply, sand
exploitation, transportation, and many other livelihoods. In many cases, water
shortages downstream also increases water pollution and salinity intrusion level
(chapter 2). Meanwhile, other groups of people benefit from hydroelectricity but do
not bear any environmental and social costs due to hydropower dam construction,
such as urban electricity users, service companies in cities, factories, resorts, and
industrial zones. They improve their living conditions by using sufficient electricity,
whereas resettled communities must struggle with negative impacts caused by
hydropower dam construction to survive from day to day. The case of Binh Dien
hydropower dam shows that the inequality in the same community is also a big issue
of hydropower development: some have more advantages to experience upward
development, whereas many other are disadvantaged and move downward in term of
development (see chapter 5). Thus, we can see that the development of this group is
not parallel with that of other groups, and the growth of the majority is at the expense
of the minority. There is no equitable mechanism to ensure fair distribution of
benefits created from hydropower development in Vietnam.

***Finally, inequity between generations: needs of current generation
versus opportunities for next generation.*** The case of hydropower development
also shows that the younger generation of displaced communities often face many
challenges in making a living because most displacement and resettlement programs
only concentrate on addressing temporary needs for current generation but not for the
next generation. In many resettlement villages, children have to drop out of school to
work in the field to help their parents. They also have to migrate to large urban
centres to earn money due to a lack of job opportunities in resettlements. Their
parents do not have sufficient land to continue cultivating and planting forests, and,
therefore, they cannot sustain farming livelihoods for their children. More
importantly, the deterioration of natural forests in both previous villages and current

resettlement villages reduces opportunities for younger generations to continue their customary livelihoods and cultural activities. Young people in many resettlement villages have to work temporarily for others, such as for forest owners in other places, to earn money for subsistence. Although they might access better education and health care services, in reality their daily difficulties limit their practical benefits from those services (chapter 4 and 5). Generally, the younger generation in the resettlement areas face great hardships in finding employment opportunities in the future due to lack of land, deterioration of natural resources, and lack of support from government, private sectors, customers of hydropower companies, and others who benefit from hydroelectricity production. The challenges of displaced communities are transposed to the next generation of resettled communities (See Figure 7-1).

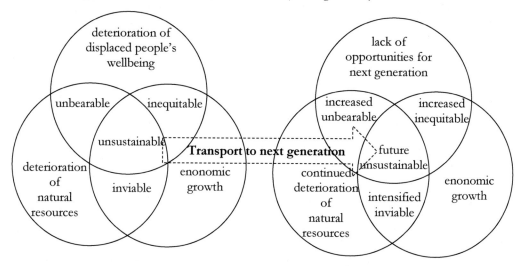

Figure 7-1 Dynamics of sustainable development dilemmas due to hydropower dam construction *Source: author's research*

Thus, four identified dilemmas of hydropower development demonstrate that the targets of sustainable development of Vietnam are being hindered. These dilemmas are dynamic and could be transposed from generation to generation to become enduring dilemmas of displaced people in Vietnam. As Cernea (2009) conjectured, development-induced displacement and resettlement is one of the major social pathologies inevitably resulting from numerous development projects and an enduring issue of development. Hydropower development-induced displacement and resettlement is painful and is difficult to cure.

7.2   Power difference versus good governance

There are many reasons leading to the current dilemmas of hydropower development in Vietnam. Our research findings show that ineffective and poor governance of hydropower development is the root cause of current dilemmas, and the large gap of power between decision-makers (government, local authorities, and hydropower developers) and affected people (displaced people and others indirectly affected) is the main reason that affected people cannot make demands of decision-makers for their legitimate interests and hold them more responsible and accountable.

In Vietnam, three main stakeholders, including government and lower authorities, state-owned and private companies, and affected people are involved in making decisions on hydropower dam construction and resettlement. The government and local authorities have potentially conflicting roles between ensuring electricity security for the country and in developing and protecting the environment and the legitimate rights and interests of affected people, especially who are displaced to make way for hydropower development. Therefore, the policy of the government is "to ensure displaced people to have at least similar living conditions or better than before displacement." Our findings show that the government has successfully fulfilled the first role, but the second role has been poorly implemented, as reflected in four dilemmas. The government of Vietnam set a very high expectation for hydropower development as an urgent need to speed up industrialization and modernization to 2020. Therefore, the government gives a higher priority to hydropower developers rather than to benefits to displaced people. Based on this premise, hydropower dam construction laws and regulations are designed to favour investors, including state-owned and private hydropower companies. The government and local authorities expropriate land from land users with the compulsory mechanism and allocate it to investors. With this mechanism, the government and local authorities do not need to negotiate with land losers to ask their consensus for land acquisition. In addition, the government and local authorities minimize the cost for investors by imposing a low compensation price framework for the loss of land and other properties to give incentives for investors. With the fixed price framework, land losers do not have any chance to negotiate with the government, local authorities, or investors about price in compensating their losses; they must accept the compensation price that the government and local authorities had pre-decided. If they refuse the decision of the government and local authorities, they are still forced to leave their lands and homes. Additionally, customary lands, such as land for Lồ Ô bamboo plantation and slash-and-burn models, are not entitled to be compensated chapters 3 and 4).

166

Thus, it can be seen that the legal framework is set for the interest of the government and hydropower developers, and provides less protection for the rights and benefits of land losers (displaced people). Often, the government and local authorities exercise the intensively top-down decision-making process and concentric governance mechanism for hydropower development, where the state is the centre of political power and authority and not displaced people (Rhodes, 1997; Pierre, 2000; Kooiman, 2003). Large-scale hydropower dams are planned and decided by the National Assembly, the government, ministries, and Vietnam Electricity Group (EVN), and there is limited consultation with other stakeholders, such as provincial and district authorities, civil society organizations, and directly or indirectly affected people. The Ministry of Industry and Trading (MOIT) often collaborate with the Provincial Department of Industry and Trade (DoIT) in planning for medium and small-scale hydropower dam construction in provinces. The Provincial People's Committee (PPC) retain final decision on granting investment licences to hydropower investors but take less consideration for the consultation of district and commune authorities, local communities, and other relevant stakeholders, such as research institutes, local mass organizations, and local NGOs. Information about descriptions of hydropower dams and decisions on compensation, displacement, and resettlement at the planning phase is provided to public inadequately. The local authorities provide this information to the public only after land acquisition decisions are made and hydropower dams are already being constructed on the ground. Furthermore, most resettlement sites are pre-determined by district and provincial authorities and hydropower investors but lack consultation with displaced communities. Therefore, displaced people's voices do not have any influence on land acquisition, compensation, displacement, and resettlement planning of the government and local authorities (See chapters 4,5and 6).

Thus, we can see that the governance of hydropower development in Vietnam is still poor and not effective enough to protect the legitimate rights and interests of affected people who must sacrifice their tangible and intangible properties for public development, and who receive little compensation and support from hydropower developers and government. This result is consistent with to the evaluation of many international organizations on the governance of Vietnam (See Figure 7-2), where the voice of the people is not respected and where the accountability and responsibility of the government to affected people is not adequate. The government and local authorities do not make the best use of resources, including government budget and natural resources, to meet the need of all people. Corruption is also a big problem of Vietnam that constrains the fair distribution of development opportunities to all people.

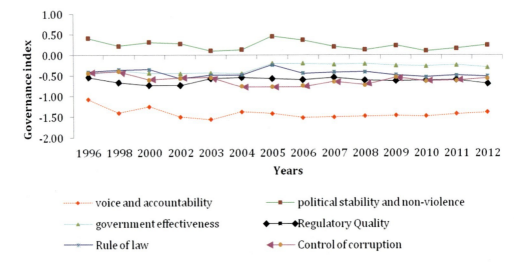

Figure 7-2 Governance indicators of Vietnam between 1996 and 2012. *Source: http://info.worldbank.org/governance/wgi/*

Power, then, is concentrated mostly in the hands of the government, local authorities, and hydropower developers, especially state-owned companies, as they are part of the government. Therefore, state-owned companies, such as EVN, have great power in deciding how hydropower is developed and power to control electricity market and services. Private hydropower enterprises, with financial strengths and good relations with government and local authority leaders, are also able to lobby policies to construct medium and small-scale hydropower dams. They receive many incentives from government and local authorities for hydropower dam construction, including low price of land lease, low compensation price frameworks, and many other tax reduction policies for private enterprises. The opaque agenda and vested interests in this coalition often strengthen the power of private companies in Vietnamese society today. Private hydropower companies and leaders of government and local authorities often have a very close interaction. They are currently considered as "*a strong interest group – nhóm lợi ích*" that is closely associated in many cases of corruption (See chapter 4). In contrast, displaced people who are not well protected by legal framework do not have much power to insist on their interests in practice. They have rights to complaint and denounce but only after all decisions had been implemented. In practice, their complaints and grievances about compensation and land are not adequately addressed because local authorities are both decision-makers and those who also resolve these complaints. In the meantime, mass organizations, who are legitimate representatives for displaced people, such as ethnicity committees, unions of farmer, women, youth, and other professional associations, are rarely invited to participate in discussing, consulting, and implementing policies of land acquisition, compensation, displacement, and resettlement. They are also bodies of government and local authorities, and, therefore, they must follow decisions made by their leaders. The case of hydropower development in Vietnam also reflects the dilemma of role conflict and ambiguity of the government of Vietnam. The government plays double roles to promote the hydropower development and to protect legitimate rights and benefits of affected people, but they do not have enough capacity to balance or do not want to balance their vested benefits when supporting hydropower developers.

This reality raises big questions: *Why does the government of Vietnam still execute the top-down decision making mechanism exclusively and hesitates to change to a better governance mechanism—bottom-up or participatory approach—for large-scale development projects? The government has the legitimacy to ensure the interests of all people, but they cannot ensure t in practice?* These questions can be explained as follows: first of all, the top-down approach has been used within the political system of Vietnam for a long time during the subsidized, bureaucratic, and administrative economy and management system between 1945 and 1986. Although many significant changes have been made since 1986, the top-down or commanding approach has deeply penetrated into generations of Vietnamese leaders, and it has become as a culture of management of Vietnamese leaders. Young leaders also have been educated to continue from their predecessors with the same principles and ideology, and the majority of them often enrol in state administrative courses that mainly embed the top-down and administrative decision-making model. They rarely learn the bottom-up, participatory model and good governance principles, or they have few chances to learn international principles, guidelines, and norms of hydropower development, such as those of World Commission on Dams (2000) and International Energy Agency (1999).

Secondly, the top-down decision-making system has enabled the government and local authorities to reduce costs for lengthy negotiation with a large number of stakeholders. For Vietnamese leaders, the leading perspective always follows "determined or decisive principles." It means that they typically make quick decisions when necessary, and they only take actions to mitigate negative impacts occurring after implementation. This type of decision-making often lacks results in preventive measures before acting. As a result of this process, most hydropower dams, for example, have been planned, constructed, and put into operation on time and even earlier than predetermined schedules. The top-down decision making mechanism has helped the government and local authorities implement hydropower projects effectively and efficiently in term of cost-benefit analysis for investors and the government (see chapter 2).

Last but not least, the top-down approach is easier for private companies and leaders of local authorities to make decisions on hydropower dam construction because local authority leaders often support private hydropower companies so as to make their investment possible and force land losers to accept the investment with strict regulations. With this bias, private companies become "delegates of local authorities" in the field and force land losers to accept compensation prices and resettlement packages without negotiation. Displaced people, as the powerless, cannot oppose to the strong alliance of power holders (See chapters 4 and 5).

The problems caused by power imbalance suggests that it is crucial to study the power network of each country to understand how power is manipulated by laws and in practice in order to explain why good governance principles are not effectively introduced into each country on the ground. Power and interests are closely intertwined, and therefore power holders often make decisions to benefit them and their allies but ignore the interests of the powerless. In a single-party state like Vietnam, too much power is given to government leaders and local authorities without effective controlling measures. This often leads to power abuse, and, consequently, ordinary people, especially evicted people, lose their power on the ground compared to their legitimate rights regulated by laws.

Power is very important for Vietnamese people because the majority strongly believes that benefits always goes with power; the more power, the more benefit. Becoming a leader is the main way to show the success of a person. The influence of power and the success of people also can be explained by the Vietnamese proverb as follows: "First of all social relation, secondly power status, thirdly finance, and lastly good policy – *nhất thân, nhì thế, tam tiền, và tứ chế*" (Vietnamese proverb).This means that the success of a person depends on four main factors, including social relation, power status, finance, and policy, where social relations are the most important factors. For example, for those with wealthy and powerful parents who are the leaders of government and local authorities, they definitely have better life than others without these conditions. In the case of hydropower development, state-owned hydropower companies, such as EVN, have very good parents—government and local authorities— who always provide them with financial and technical supports to advance hydropower investment. Other companies, who do not have good parents, should have a good relationship with a political, economic, social, and professional network, especially good relationship with "god parents," such as leaders of the government, local authorities, companies, or other institutions. From these, they can have good opportunities to change their fate by lobbying their interests with their "god parents." In the case of hydropower development, many private hydropower companies established good channels to lobby hydropower investment policies.

The third important factor for success: if people do not have good parents or god parents, money is key. For example, many large companies with abundant finance were invited to invest in hydropower development since the government and state-owned enterprises did not have sufficient finances to build all hydropower dam as planned.

The fourth factor is good policy. If people do not have supports of the previous factors, they must make good use of policy. For instance, most people displaced due to hydropower dam construction are ethnic minority who have neither good parents nor god parents or money. They are mostly poor farmers depending very much on the policies of the government and local authorities to advance their livelihood success after resettlement. However, policies on compensation and resettlement are still too weak to help displaced people overcome difficulties caused by hydropower dam construction. Additionally, displaced people do not have enough political power to insist the government and local authorities on their legitimate rights and benefits. The findings described in chapter 5 show that living conditions of powerless households' become worse than other households of village leaders. It is clear that lack of power balance has limited the legitimate rights and benefits of displaced people from the development of hydropower dams in Vietnam. The Party-State declares that leaders and staffs of the Party-State are loyal servants—*đầy tớ trung thành*—of the entire people, but the case of hydropower development shows that the motto does not work in practice because displaced people always must insist upon leaders and staffs of the Party-State for their legitimate rights and interests within the process of land acquisition, compensation, displacement, resettlement, and livelihood support programs. Therefore, the motto should be changed to "leaders and staffs of the government and local authorities are parents of the entire people." The government must play as "a good and fair mediator or referee" of the game to treat various teams and players fairly and accountably. Only if parents treat all children fairly, including state-owned enterprises, private companies, customers of hydropower, and affected people (especially those displaced as a result of dams), equitable development could be attainable.

To date, hydropower has become a controversial issue, and displaced communities have become the centre of attention of researchers, journalists, environmental and social activists, international organizations, financial sponsors, international NGOs, Vietnamese NGOs, inter-governmental organizations, and even the government of Vietnam. Resettlement villages become study sites, and displaced communities become the targets of rural development projects and beneficiaries of financial aids and humanitarian relief. Many organizations are involved in improving the outcomes of hydropower development and resettlement that create new opportunities for better governance of hydropower development and improved livelihoods for displaced communities.

7.3   New trends and opportunities for better governance and sustainable hydropower development

The first opportunity is that Vietnam has carried out profound legal framework reforms in recent years. The new Constitution 2013 provides new regulations to secure civil rights and strengthen the responsibility of the Party to the entire people, in which it emphasises that the Party must be subject to supervision by the people. If they make wrong decisions, negatively affecting the interests of nation and the people, the Party shall be responsible to their decisions. This implies that the Party must enhance their responsiveness and accountability to the people. This is a new regulation, as this was not stated clearly in the previous Constitution, when all development policies are first made based on the resolutions of the Party, and people did not have any rights to question the decisions of the Party. Additionally, the new Constitution defines clearly the boundary of legislature, judiciary, and enforcement departments that used to be blurred in the former Constitution. This expects to enhance the efficiency and effectiveness of grievance and complaint resolution incurring from land acquisition for hydropower dam construction because it stop exercising the "double-roles playing" of government and local authorities in making and implementing their decisions, as well as in resolving grievances at the same time.

Following the new Constitution, the new Land Law was also amended in 2013. The new land law is the main legal framework for land acquisition, compensation, displacement and resettlement, and livelihood support. It aims to recognize more land use rights and to provide land losers with high compensation and participation in decision-making process. The process of land acquisition is expected to be more transparent. Decision makers are requested to be more accountable and responsive (See chapter 3). Furthermore, more regulations have been issue to recognize and encourage Vietnamese NGOs to become involved in decision-making regarding social-economic development, policy review, and supervision (See chapter 6). It reflects that the party-state has made many legal improvements to promote elements of good governance into practice. New regulations have enabled VNGOs to have more autonomy to criticize the decisions of government and local authorities on hydropower dam construction and to review plans of dam construction, as well as to organize campaigns in order to hold government and local authorities more accountable to problems of hydropower development and resettlement. This also has strengthened the influence of VNGOs and increased the voice of displaced people to government and local authorities.

On account of the intervention of VNGOs, government and local authorities have become more responsible for providing displaced people with more land for agriculture production and forest plantation. Many livelihood development programs have been initiated and the government and local authorities have become more careful than before when making decisions on building new hydropower dams. Thus, the number of VNGOs is growing and playing important roles to bring good governance into practice. However, it requires establishing more VNGOs cross the country, from rural to urban areas, to become a stronger and stable network, playing as the representative of the powerless to hold government and local authorities more accountable, as well as to promote a more democratic decision-making process. Only if has Vietnam a strong and stable civil society with a comparable voice to the government and local authorities will the costs and benefits of hydropower be shared equally to all stakeholders, not only have displaced people who have borne the costs to date.

The second opportunity is that the government have formulated the first benefit sharing policy based on the technical and financial supports of several international NGOs. In which, the model of forest environment service payment (FPES) is quite successful in collecting money from hydropower companies to support displaced people and forest owners involving in forest protection. So far, many displaced households have benefited from this model (see chapter 6). This success also shows a positive partnership among the government, local authorities, and international NGOs. The constructive and collaborative interaction among them has produced a good "benefit sharing model" to facilitate the improvement of hydropower development.

The third opportunity is that the growth of industry and service sector in Vietnam has indirectly contributed significantly to the improvement of the living conditions and livelihoods of several displaced communities living near urban areas. This is shown in the case study of chapter 5, in which the displaced households of Bo Hon village recovered their income moderately after 8 years of resettlement. Many earned a significant income when they became involved in planting Acacia forest and working for textile factories, as both industries have been growing considerably in Vietnam (See chapter 5). However, these improvements have taken place in small-scale areas, and, therefore, we need further study to expand this resettlement model that relocates displaced people to near urban regions where they are able to access better job opportunities to improve their living conditions.

7.4   Revisit the global debate on hydropower development, land and water grab

There is no smoke without fire – Không có lửa làm sao có khói (in Vietnamese)

From an analysis of development dilemmas, power differentiation, and adverse consequences in Vietnam, I strongly believe that there is a justifiable reason for opponents to carry out campaigns to ask the government and hydropower developers to stop building hydropower dams worldwide. The adverse impacts of hydropower dams on society and environment have increased significantly in uncertain magnitudes and degrees, such as earthquakes, dam failure, and conflicts over access to natural resources. With ambitious hydropower dams building projects that include hundreds of dams, Vietnam has been paying the price for deterioration of natural resources and a rise in social problems due to hydropower dam construction that will take many decades and huge costs to mitigate the dams 'negative impacts. Although, the government of Vietnam has declared a stop to large hydropower dam building in the coming years, hundreds of medium and small hydropower dams have still been approved to begin construction across the country. Meanwhile, the facts in Vietnam show that small hydropower dams also cause huge impacts on environment and society because most small dams are built by private companies whose ultimate goal is to generate profits for their companies, while taking less consideration for the protection of environment and improvement of displaced people's living conditions. Therefore, it is necessary to study the issues of small hydropower dams in the future, as well.

When it comes to the global debate on land grabbing, the description of hydropower development in Vietnam reflects that land acquisition for dam construction is often legal, but it does not mean that its process is fair for all stakeholders. Large-scale land acquisition for hydropower dams in Vietnam embeds much characteristics of domestic land grab led by the government, local authorities, and its development counterparts. However, the current debate on global land grabbing pays less attention to this process (See chapter 3 and Mayke & Annelids, 2014). Vietnam has been expanding hydropower investment other countries, including Laos and Cambodia. This also may transfer the same model of development and its impacts to environment and society in the destination countries. Therefore, it is crucial to understand how Vietnam executes their hydropower dams in Laos and Cambodia to provide more knowledge for global land grab debates. Furthermore, the issue of water grabbing related to hydropower dam construction is also crucial to consider because the current debate focuses much on land grabbing or large-scale land acquisition rather than water grabbing.

The development of hydropower in Vietnam also indicates a paradox between the governance system of less-developed countries and international standards and guidelines on good governance. While less-developed countries, like Vietnam, exercise a top-down decision-making process and monocentric governance to favour their investment that does not require public acceptance, most international guidelines and standards requires decision-makers and hydropower developers to carry out bottom-up, participatory, and multi-stakeholder approaches to attain a complete consensus with all relevant stakeholders at different levels for making final decisions on building hydropower dams, such as the decision making framework of WCD (2000). This approach often delays the investment, and many leaders of the government and local authorities do not accept it (Biswas, 2012). For example, in several conferences, Vietnam leaders often think that Vietnam is still a poor country, and, therefore, it is inevitable to promote economic development without accepting costs of environment and society. As a result, leaders of Vietnam often make decisions quickly and decisively to seize the opportunity to develop; if not, they feel the country may lag behind other countries (Author's note at conferences[29]). This perception has resulted in a weak sustainability approach, which often leads to weak sustainable development. This ideology often hinders leaders of Vietnam to accept the high standards of international organizations, like WCD guidelines and standards (2000). Therefore, to make a change, first of all, it is crucial to educate these leaders to be more aware of equitable and sustainable development, then to provide them technical and financial supports to enhance the capacity of good governance to transfer knowledge to action. It is similar to the intervention of Winrock International in Vietnam to promote a successful model of forest environment service payment model as benefit sharing model for displaced communities. Furthermore, repeated training and sustaining action requires a long-term commitment between international organizations and government of less-development countries. Cultural difference, particularly the culture of power relation, should be taken into account to facilitate capacity building and implement actions on the ground for each country and localities. I strongly agree with Nelson Mandela that "Education is the most powerful weapon which we can use to change the world" (Nelson Mandela, Speech to the Mindset Network, July 16th, 2003).

In the end, we have witnessed that issues of hydropower are not new to us anymore; the benefits and costs of hydropower has been clearly demonstrated on a global scale. The solutions to ensure sustainable hydropower development have been

---

[29] Three conferences in Vietnam: (1) Annual conference of Vietnam River Network in 2012; (2) The conference of Department of Irrigation; and (3) Participation of media, communities, and civil society organizations in facilitating transparency, consultation, and environmental supervision in development projects in Da Nang city, Vietnam in 2014.

widely publicized and most countries in the world are able to access those guidelines. Good practices of hydropower dam construction and resettlement programs are presented worldwide. However, the main question is how hydropower dams should be best planned, built, operated in less or under-developed countries where hydropower dam construction is an inevitable option to satisfy the electricity shortage and ensure basic needs for poor people. In this case, elements of good governance should be included, in which public acceptance and adequate opportunities for affected people to be consulted, discussed, negotiated, and benefited must be incorporated in the making-decision process and implementation of hydropower projects.

For Vietnam, the leader of EVN declared that the country has sufficient electricity for whole country with the significant support of hundreds of hydropower dams and will stop building new, large hydropower dams by 2017. But hundreds of medium and small hydropower dams are being constructed and are planned to be constructed in the coming years. This study shows that all types of hydropower dams cause tremendously environmental and social costs that are a huge burden for the country currently and for the next generation. Therefore, I strongly believe that it is not wise to construct more hydropower dams and add more burdens to the next generation and environment. There is no reason for the government to promote building more hydropower dams, even medium or small dams, because the loading capacity of ecosystem and population does not allow, making it worse. Henceforth, the most important role of government, hydropower developers, and all society is to search for innovative solutions and put more effort and money to solving the existing consequences of the hydropower dams already constructed. We do not have time to hesitate because we have borrowed too much from the nature and displaced people and have not fully compensated for losses. If we continue to borrow, we will not have the ability to pay back. As a result, the human and natural ecosystem will be disrupted. I strongly believe that returning what we borrow is also an ethical issue to pursue in equitable and sustainable development.

References

Cernea, M. (2009). Introduction: Resettlement – an enduring issue in development. *The Asia Pacific Journal of Anthropology, 10*(4), 263-265. doi:10.1080/14442210903079756

Kooiman, J. (2004). Governing as governance. *International Public Management Journal, 7*(3), 439-442.

MOIT, 2013. The removal of hydropower projects is negligible. http://www.moit.gov.vn/vn/pages/Tinchuyende.aspx?MachuyendeNCT_CT &IDNews=2451

Peters, B. G., & Pierre, J. (2000). Governance, politics and the state. Palgrave Macmillan: London, UK.

Rhodes, R. A. (1997). *Understanding governance: Policy networks, governance, reflexivity and accountability.* Open University Press.

APPENDIX 1. QUESTIONNAIRES FOR HOUSEHOLD SURVEYS FOR TWO CASE STUDIES IN A LUOI AND BINH DIEN HYDROPOWER DAMS

Questionnaire number:
Date of interview:
Interviewer:

Component I: General information of displaced households

1. Name of household head: 2. Sex: a. Male   b. Female
3. Age: 4. Number of people:
5. Religion:
6. Ethnicity:
      a. Kinh
      b. Ta Oi
      c. Ka Tu
      d. Pa Co
      e. Van Kieu
      f. Other, namely
7. Status of marriage   a. Couple   b. Wisdom   c. Not married
8. Education: a. Primary   b. Secondary   c. High school
                         d. College/University   e. illiteracy
9. Details of households

Members of family	Name	Age	Sex	Marriage status		Social position	Health status	Education	Main occupation	
				Yes	No				Before displacement	After resettlement
Husband										
Wife										
Children										
Other members										

Component II: Land appropriation, displacement and resettlement process

10. Village before displacement:
11. Name of resettlement site:

12. How do you know about the dam construction planning?
        a. Completely informed
        b. Half informed
        c. Completely did not inform

13. When were you informed about the appropriation of your land to the dam construction (dd/mm/yy)?

14. Did you understand why your land was appropriated?
        a. Completely
        b. Partly
        c. Not at all

15. Were you informed about the displacement and resettlement prior to this event?
        a. yes (if yes, go to question 16 - 19)
        b. no

16. Did you have sufficient information about the displacement and resettlement?
        a. Completely
        b. Partly
        c. Not at all

17. Who informed you about the land acquisition, displacement and resettlement process?
        a. Village leaders
        b. Commune leaders
        c. District leaders
        d. Provincial leaders
        e. Hydropower company
        f. From other villagers, neighbours, friends, and relatives
        g. Other sources: ....

18. What did you know through these sources about the land acquisition, displacement, and resettlement process?
....
....
....

19. How did the process on choosing the resettlement?
        a. All villagers of the community chose their resettlement sites independently
        b. Hydropower company negotiated with the community
        c. Local governments together with villagers selected the resettlement site
        d. Hydropower company and local government discussed with the community
        e. No discussion and negotiation

20. Did you attend any meeting organized by local authorities and hydropower company?
        a. Yes (if yes, go to question 21, 22, 23)
        b. No

21. Were your 'resettlement' wishes/needs/demands taken into account?

       a. Completely

       b. Partly

       c. Not at all

22. Were you informed and consulted about the negative/positive impacts due to displacement and resettlement?

       a. Completely

       b. Partly

       c. Not at all

23. Did you give your consent regarding the resettlement process?

       a. yes

       b. no

24. When did you move to this village?

       a. Just when the project started

       b. Six months before the project started

       c. One year before the project started

       d. More than one year after the project started. Explain

       why? ....

       ....

25. What is your opinion about the resettlement process?

       a. Very positive

       b. Positive

       c. Neutral

       d. Negative

       e. Very negative

       f. No opinion

Component III: compensation and assistance process

26. Have you received any form of compensation and assistance?

       a. yes

       b. no (go to question …)

27. How were you compensated?

       a. Money....................VND

       b. Land.................... ha

       c. Houses.................. m²

       c. Trainings,

namely..

       d. Other assistances,

namely..

28. Have you received any consultation to spend your compensated money and land?

       a. yes

       b. no

29. What did you spend compensation money for?

       a. Housing.. VND

       b. Furniture.. VND

       c. Savings.. VND

       d. Daily expenditures.....................................VND

       e. Land..VND

       f. Others, namely... VND

30. How did you know about your land and property losses due to hydropower dam construction?

       a. You claimed yourself

       b. Local governments estimated

       c. Hydropower company and local governments surveyed with you

       d. Other sources: please

explain..

31. How did you satisfy with the list and quantity of your land and property losses estimated for compensation?

       a. totally acceptable   b. partly acceptable

       c. neutral   d. very inadequately

       e. inadequately   f. no opinion

32. Did you have any negotiation with local governments and Hydropower Company about the price for compensation?

       a. yes

       b. no

33. How did you satisfy with the price of compensation?

       a. very satisfactory   b. satisfactory

       c. neutral   d. very disappointed

       e. disappointed   f. no opinion

34. What do you think about the assistance of hydropower company and local authorities?

       a. Very positive

       b. Positive

       c. Neutral

       d. Negative

       e. Very negative

       f. No opinion

Component IV: Risks and vulnerabilities of displaced households

4.1. Landlessness

No	Questions	Before displacement	After resettlement
35	How much land do you have?	……... ha	……... ha
36	What is the main type of land use of your land? (more answers possible)	a. Agriculture …….......ha b. Forestry ……...........ha c. Garden …....…........ha d. Other(s), namely.... .... ha	a. Agriculture …….......ha b. Forestry ……...........ha c. Garden …....…........ha d. Other(s), namely.... .... ha
37	What kind of crops do you plant on your agricultural land? (more answers possible)	a. Cassava b. Wet paddy c. Dry paddy d. Rubber e. Other(s), namely	a Cassava b. Wet paddy c. Dry paddy d. Rubber e. Other(s), namely
38	What kind of crops do you plant in your garden? (more answers possible)	a. Fruit b. Grass c. Herb d. Other(s), namely ……... ..	a. Fruit b. Grass c. Herb d. Other(s), namely ……...
39	Do you sell your crops or use it for own consumption?	a. Selling crops b. Using crops for own consumption c. Both	a. Selling crops b. Using crops for own consumption c. Both
40	How is the condition of your land?	a. Very good b. Good c. Not bad/not good d. Bad e. Very bad	a. Very good b. Good c. Not bad/not good d. Bad e. Very bad
41	Do you use fertilizer for your land? (more answers possible)	a. Yes, natural fertilizer .... VND b. Yes, chemical fertilizer .... VND c. No, nothing	a. Yes, natural fertilizer.................. VND b. Yes, chemical fertilizer .... VND c. No, nothing
42	Is your land registered?	a. Yes b. No	a. Yes b. No

4.2. Joblessness

No	Questions	Before displacement	After resettlement
43	Average monthly household income (VND)	....	....
44	Main household sources of income (per month on average in VND)	a. Agriculture .... b. Garden .... c. Forestry..................... d. Fishing .... e. Livestock .... f. Construction g. Small scale business h. Other(s), namely	a. Agriculture .... b. Garden .... c. Forestry..................... d. Fishing .... e. Livestock .... f. Construction g. Small scale business h. Other(s), namely
45	Temporality of formal employment contract (if applicable)	a. 1 - 6 months b. 7 - 11 months c. 1 - 2 years d. More than 2 years e. Unknown	a. 1 - 6 months b. 7 - 11 months c. 1 - 2 years d. More than 2 years e. Unknown
46	Have one or more of the household members migrated to support the household?	a. Yes, namely .... b. No (continue with question ?)	a. Yes, namely .... b. No (continue with question ?)
47	How much does the household receive per month from remittances (VND)?	....	....

48. Due to displacement and resettlement, which main economic activities did you lose?
....
....

49. Has the process of resettlement led to a loss of income sources?
     a. yes
     b. no

4.3. Food insecurity

No	Questions	Before displacement	After resettlement
50	How much food crops do you produce (kg per year)?	.... kg per year	.... kg per year
51	How much food crops do you produce for own use (kg per year)?	.... kg per year	.... kg per year

52	Does the food crop production satisfy own nutritional needs?	a. Yes b. Partly c. Not at all	a. Yes b. Partly c. Not at all
53	How much of the households monthly income is spent on food (VND)?	....	....

54. Did your household experience any diseases after resettlement which was absent or less intensive before resettlement? If so, which one(s)?

....

....

....

....

4.4. Loss access to common pool resources

No	Questions	Before displacement	After resettlement
55	Do you have access to common property?	a. Yes b. No	a. Yes b. No
56	What type(s) of common property do you use?	a. Forest b. River c. Agricultural land Other(s), namely	a. Forest b. River c. Agricultural land Other(s), namely
57	What kind of activities is the common property used for? (more answers possible)	a. Recreation b. Education c. Income generating activities, namely d. Other(s), namely	a. Recreation b. Education c. Income generating activities, namely d. Other(s), namely
58	How much do you earn from these activities per month (VND)?	....	....

59. Are you satisfied with current common property resources?

        a. Completely

        b. Partly

        c. Not at all,

because..

4.5. Loss access to public services

No	Questions	Before displacement	After resettlement
60	Does your household have access to clean drinking water?	a. Yes b. No	a. Yes b. No
61	Does your household have access to	a. Yes	a. Yes

184

	sanitation facilities?	b. No	b. No
62	To what extent are you satisfied with the available sanitation facilities?	a. Very positive b. Positive c. Neutral d. Negative e. Very negative f. No opinion	a. Very positive b. Positive c. Neutral d. Negative e. Very negative f. No opinion
63	Does your household have access to electricity?	a. Yes b. No	a. Yes b. No
64	To what extent are you satisfied with the electricity facilities?	a. Very positive b. Positive c. Neutral d. Negative e. Very negative f. No opinion	a. Very positive b. Positive c. Neutral d. Negative e. Very negative f. No opinion
65	Do you have access to medical services?	a. Yes b. No	a. Yes b. No
66	To what extent are you satisfied with the available medical services?	a. Very positive b. Positive c. Neutral d. Negative e. Very negative f. No opinion	a. Very positive b. Positive c. Neutral d. Negative e. Very negative f. No opinion
67	Do your children go to primary school? (if applicable)	a. Yes, ... children b. No, because.............	a. Yes, ... children b. No, because.............
68	To what extent are you satisfied with the primary educational services? (if applicable)	a. Very positive b. Positive c. Neutral d. Negative e. Very negative f. No opinion	a. Very positive b. Positive c. Neutral d. Negative e. Very negative f. No opinion
69	Do your children go to secondary school? (if applicable)	a. Yes, ... children b. No, because.............	a. Yes, ... children b. No, because.............
70	To what extent are you satisfied with the secondary educational services? (if applicable)	a. Very positive b. Positive c. Neutral d. Negative e. Very negative f. No opinion	a. Very positive b. Positive c. Neutral d. Negative e. Very negative f. No opinion

4.6 Social disarticulation

Questions	Before displacement	After resettlement

71. What is your religion?		
72. Did you participate in buffalo killing festival?	a. yes b. no	a. yes b. no
72. Did you practice any new rice ceremony?	a. yes b. no	a. yes b. no
73. Food processing?	a. Traditional b. Follow Kinh people's style c. Others:	a. Traditional b. Follow Kinh people's style c. Others:
74. Wearing?	a. Traditional b. Follow Kinh people's style c. Others:	a. Traditional b. Follow Kinh people's style c. Others:
75. Marriage ceremony?	a. Traditional b. Follow Kinh people's style c. Others:	a. Traditional b. Follow Kinh people's style c. Others:
76. Ancestor worship?	a. Traditional b. Follow Kinh people's style c. Others:	a. Traditional b. Follow Kinh people's style c. Others:
77. Giving name for children?	a. Traditional b. Follow Kinh people's style c. Others:	a. Traditional b. Follow Kinh people's style c. Others:
78. Weaving traditional clothes?	a. yes b. no	a. yes b. no
79. Language?	a. Your traditional language b. Kinh (Vietnamese) c. others:	a. Your traditional language b. Kinh (Vietnamese) c. others:
80. Hunting?	a. yes b. no	a. yes b. no
81. Did you children still practice your traditional culture?	a. yes b. no	a. yes b. no
82. Do you feel that your characteristics are similar to Kinh people?	a. completely b. very similar c. quite similar d. not clear e. completely not	a. completely b. very similar c. quite similar d. not clear e. completely not
83. Is your custom change due to living near Kinh people after resettlement?		a. yes b. no
84. Is your resettlement site close to urban areas?		a. yes (if yes, answer question 85) b. no
85. If living near cities or urban areas, how does		a. very influential b. quite influential

186

that influence your culture?		c. partly influential d. not clear e. not at all
86. Do you have any persons who are leaders of village, commune, and other authority level?	a. yes b. no	a. yes b. no
87. Did your family help others during difficult periods?	a. yes b. no	a. yes b. no
88. Are you able to borrow rice or receive help from other families?	a. yes b. no	a. yes b. no
89. Do you have labour exchange with other households?	a. yes b. no	a. yes b. no
90. Do you have mutual help during wedding or funeral events?	a. yes b. no	a. yes b. no
91. Do you go to see ill people in your village?	a. yes b. no	a. yes b. no
92. Do you ask your neighbours to look after your houses when you are away for working?	a. yes b. no	a. yes b. no
93. How do your relatives help you in the field for agriculture production?	a. very often b. quite often c. sometimes d. not at all	a. very often b. quite often c. sometimes d. not at all
94. Do you receive money help from your relatives living far away from your village?	a. yes b. no	a. yes b. no
95. Are you a member of any association in your village, commune or outside?	a. yes b. no	a. yes b. no
96. Do you receive any preferential policies, such as policy for the poor, disabled, or veteran?	a. yes b. no	a. yes b. no
97. Have you ever	a. yes b. no	a. yes b. no

received any support from NGOs?		
98. Do you have many friends in cities?	a. yes b. no	a. yes b. no
99. Do you borrow money from banks?	a. yes b. no	a. yes b. no
100. In general, how do you see the relationship among households in your village?		a. very good b. quite good c. neutral d. quite bad e. very bad f. no opinion

Component V: Important livelihood adaptation strategies after resettlement

5.1. Work for others

101. How many people are in your family working for someone else?...............

102. How many are they female and male?

      a. male.................   B. female........................

103. When did they start working for others after resettlement?

      a. right moved in the resettlement site

      b. after one year   c. 1-2 years   d. 2-3 years   e. > 3 years

104. What are they doing?

      a. Work for Acacia forest owners   b. others:

....

105. How do they get to work place?

      a. on foot   b. by motor bike c. by bike   d.

others:.................................

106. How often do they work for others?

      a. very often   b. quite often   c. rarely

107. How many days a year do they work for others (days)?...

108. How much do they earn per day? ....VND

109. How does your family spend earned money for?

      a. buying rice   b. daily foods   c. for schooling d. health care   e. buying furniture

      e. bike, motor bike   f. mobile phone g. buying land   h. livestock

      i. acacia plantation   j. for migration k.

others:...

110. How do you see the demand for this kind of work?

      a. very demanding   b. quite demanding   c. decreasing very significantly

d. decreasing quite significantly   e. not clear

5.2. Migration to big cities for working

111. How many people are in your family migrating to cities?................

112. How many are they female and male?

      a. male................   b. female........................

113. When did they start migrating after resettlement?

      a. right moved in the resettlement site

      b. after one year   c. 1-2 years   d. 2-3 years   e. > 3 years

114. What are they doing?

      a. textile factory   b. small business   c. for state companies   d. others: ....

115. Which city are they living now?...

116. How often do they stay in those cities?

      a. very often   b. quite often   c. sometimes   d. rarely

117. How many months a year do they work in cities (month)?.......................................

118. How much do they earn per month? ....VND

119. How much remittance have you received from your migrants?..................................VND

120. How does your family spend earned money for?

      a. buying rice   b. daily foods   c. for schooling   d. health care   e. buying furniture

      e. bike, motor bike   f. mobile phone   g. buying land   h. livestock

      i. acacia plantation   j. for migration

      k.others:...

121. How do you see the demand for this kind of work?

      a. very demanding   b. quite demanding   c. decreasing very significantly   d. decreasing quite significantly   e. not clear

5.3. Economic forest plantation

122. Did you have Acacia forest or other economic forest plantation before resettlement?

      a. yes, how many hectares.................................   b. no

123. Do you have Acacia forest or other economic forest plantation after resettlement?

      a. yes, how many hectares..   b. no (go to question 130)

124. When did you start planting Acacia forest after resettlement? (mm/yy).........................

125. Why did you choose Acacia forest for planting? ....

126. Where did you have land for planting Acacia forest?

      a. you bought   b. from local authorities   c.   from   NGOs' assistance   d. from hydropower company   e.   for   other sources:...

127. How old are your forests? (years)...

189

128. How many times have you sold your Acacia forests?

a. 1...............VND   b. 2...............VND

c. 3...............VND   d. 4.......VND   e. others:........................

129. Who did you sell your Acacia forests?...

Component VI: What do you evaluate your living standards after resettlement compared to those conditions before displacement?

No	Assessment fields	Much worse	Worse	Unchanged	Better	Much better
		a	b	c	d	e
130	Culture value					
131	Income					
132	Food security					
133	Access to school					
134	Access to health care					
135	Access to electricity and water					
136	Pollution situations (water, air, noise...)					
137	Transportation conditions					
138	Mutual helps					
139	Relation with local authorities					
140	Access to mass media					
141	External relation with outsiders					
142	Living conditions compared to neighbouring villages or communes					
143	Natural disasters (flood, drought, ...)					

SUMMARY

Dilemmas of sustainable hydropower development in Vietnam: from dam-induced displacement to sustainable development?

Hydropower is one of the biggest controversies in Vietnam in recent decades because of its adverse environmental and social consequences that constrain the objectives of the Vietnamese government on equitable and sustainable development, especial negative impacts on displaced people who make way for hydropower dam construction. The goal of this book is to explain the controversies related to hydropower development in Vietnam in order to make policy recommendations for equitable and sustainable development. This book focuses on the analysis of emerging issues, such as land acquisition, compensation for losses, displacement and resettlement, support for livelihood development, and benefit sharing from hydropower development. The analysis emphasizes the role of different stakeholders in the decision-making process for hydropower development in Vietnam as a means to find a better governance model. This study was conducted from 2010 to 2014; however, the data and results of previous studies are also used to explain more completely the trends of this controversial issue. Qualitative and quantitative research methods were applied for data collection and analysis of research problems. Opinion of various stakeholders at different levels were collected and analyzed to understand better the roles and influences of the parties involved in the process of making decisions in relation to hydropower dam construction in Vietnam.

Since independence in 1945, Vietnam has considered hydropower as one of the most important strategies to promote industrialization and modernization of the country. Hydropower potential in Vietnam is 20,560MW. Currently, Vietnam has installed 13,694MW, accounting for 70% of hydropower potential. According to the Electricity Development Plan, Vietnam will have completed the hydropower development plan by 2030. In 2013, hydropower accounted for 40% of total electricity production and provided sufficient electricity to Vietnam. Electricity stability has contributed to the growth of Vietnam's economy. The World Bank and the Asian Development Bank appreciate Vietnam efforts, as the electrification program is very successful. Currently, more than 90% of rural areas are connected to the national electricity network. This has contributed to promoting socio-economic development and improvements of services and infrastructure of rural areas. The success hydropower and electricity sector is a source of pride for Vietnamese people (see chapter 2).

However, the benefits of hydropower cannot be compared with the negative impacts that hydropower caused to the environment and society. The construction of hundreds of large-scale hydropower dams has caused a huge loss of natural forests and bio-diversity. Hundred thousand hectares of fertile land areas for agricultural production are submerged under hydropower reservoirs. Also, hydroelectric dams have caused many conflicts over water resources for various economic sectors; for instance, the conflict between the use of water to produce electricity and water needs of people downstream for agriculture, transportation, and drinking water. Many dams have failed, causing damage to property downstream. In particular, earthquakes have occurred near hydropower dams threatening locals and resettled people's lives. To date, approximately 200,000 People have been displaced and relocated for the construction of hydroelectric dams, of which over 90% are ethnic minorities. The majority of resettled people have no stable life after resettlement, and their living standards are increasingly more difficult than before resettlement. In practice, very few cases of resettlement due to hydropower dam construction are considered as successful examples in Vietnam (see chapter 2, 3, 4 and 5).

Consequences caused by hydropower development are defined by the following underlying causes. First of all, land acquisition policies, compensation for damages, displacement and resettlement in Vietnam have made significant improvement, but this situation is still not perfect. In addition, there is a big gap between policy and practice. The implementation of these policies at the local level is poor, the responsibility of local governments is not high, and local staffs' capacity for policy implementation is weak. Very often, policy implementation is imposed by government and local authorities. There is a lack of involvement of key stakeholders except for hydropower investors and local governments. Those people whose land is appropriated do not have sufficient rights and opportunities to participate in making decisions related to land acquisition, compensation, displacement and resettlement. Most decisions are given and implemented by local authorities and hydroelectric investors, and, therefore, the voice of displaced people is not considered as important in these decision-making processes. For example, the price of land compensation and other assets is made by the Provincial People's Committee each year; it is strictly fixed, and, therefore, land losers cannot negotiate with investors to compensate at market prices. Furthermore, displaced people do not have any opportunity to select an appropriate site for resettlement, as hydropower investors and local governments select most resettlement areas. Consequently, most of resettlement areas are located in difficult areas, with narrow land and poor soil quality.

More importantly, investors are responsible for the compensation and support for resettlement of displaced people for only 1-2 years; afterward, they do not have any responsibility to share the benefits of building hydropower dams. For

192

example, resettled people are subsidized with electricity within 1 year, but then they have to pay for electricity. This is a paradox because displaced people have sacrificed their properties and health for hydropower dam construction, but they are then not provided free electricity. In addition, investors and local governments do not have along-standing policy to support resettled people to recover and develop their livelihoods. Most resettled people do not have adequate land for production and employment and most face a more difficult life after resettlement. Consequently, they return to using outmoded and illegal slash-and-burn agriculture and destroy forests or return to the areas they once lived near the dams for forest exploitation and agriculture production, causing many conflicts between resettled people and hydropower investors, local authorities, and local people (See chapters 3, 4, and 5).

The negative impacts of hydropower on the environment and society led to a movement against hydroelectric dam construction by a network of non-governmental organizations in Vietnam and among resettled communities. This network has become more and more organized and powerful. As a result, their voice has strong influence on hydropower development policy of the Vietnamese government. In addition, NGOs have found an important position in Vietnamese society and the Vietnamese government has become more tolerant to NGOs activities. Today, NGOs are protected by formal laws. So far, NGOs have expanded their activities in different geographical areas and fields since they succeeded in anti-dam movement. Besides VNGOs, international NGOs also have participated in improving the benefit sharing policies from hydropower investment by promoting the Vietnamese government and local authorities to implement the payment for forest environment service (FPES) model for the protection of forests for hydropower reservoirs. To date, many resettled communities have benefited from this model. The success of this benefit sharing pilot has prompted the government to expand the model on a national scale; as a result, Vietnam developed the first policy of its kind for payment for forest environmental services in Southeast Asia (See chapter 6).

Thus, the study results reflect that the construction of hundreds of hydropower dams has caused many negative consequences for the environment and society, especially for displaced and resettled people. Globally, the Vietnam example shows that these problems occur in most hydroelectric dam construction, in most regions, with dams of different sizes—from large-scale to small-scale hydropower dams. These problems continue to occur and last for many years, as there is no solution to solve these problems efficiently. In other words, issues related to hydropower development, such as land acquisition, compensation for damages, displacement and resettlement, livelihood rehabilitation and development for resettled people is extremely difficult and cannot be successfully resolved. Therefore, we can see that the government of Vietnam, investors, and society do not have sufficient

193

capacity and effective solutions to solve existing problems caused by hydropower dam construction. If the government of Vietnam continues investing in hydropower development, the burden to environment and society will become more severe than it might be possible to cure. Hydropower development can be equitable and sustainable only when it produces benefits that are shared equitably to displaced people and effectively allocated to rectify the consequences that it causes to the environment and society.

SAMENVATTING

Dillema's als gevolg van de ontwikkeling van waterkracht energie: gedwongen verplaatsing versus duurzame ontwikkeling

De afgelopen decennia is er in Vietnam veel controverse ontstaan rondom het opwekken van waterkracht energie vanwege de negatieve gevolgen die dit heeft voortgebracht voor de maatschappij en het milieu. De bouw van stuwdammen heeft met name geleid tot negatieve gevolgen voor verplaatste gemeenschappen. Deze gevolgen betekenen een inperking van de doelstellingen van de Vietnamese overheid om te komen tot rechtvaardige en duurzame ontwikkeling. Het doel van dit boek is om de controverse rondom de ontwikkeling van waterkracht in Vietnam te analyseren en aanbevelingen op het gebied van beleid te geven. Dit boek richt zich op de analyse van nieuwe vraagstukken, zoals land acquisities, schadevergoedingen, verplaatsing en hervestiging, ondersteuning in het opnieuw opbouwen van levensonderhoud en de verdeling van de opbrengsten van waterkracht. De analyse benadrukt de rol van de verschillende belanghebbende partijen in het besluitvormingsproces rondom de ontwikkeling van waterkracht als middel om een beter bestuurlijk model hiervoor te ontwikkelen. Deze studie werd uitgevoerd van 2010 tot 2014 en is aangevuld met de bevindingen van eerdere studies rond dit onderwerp. Bij de dataverzameling en - analyse is gebruik gemaakt van zowel kwalitatieve als kwantitatieve onderzoeksmethoden. Hierbij zijn de meningen van diverse belanghebbenden op verschillende niveaus verzameld en geanalyseerd, om zo tot een beter begrip te komen van de rol van betrokken partijen in het besluitvormingsproces rondom de bouw van stuwdammen in Vietnam.

Sinds de onafhankelijkheid in 1945 heeft Vietnam waterkracht beschouwd als één van de belangrijkste strategieën ter bevordering van industrialisatie en modernisering van het land. Het potentieel voor waterkracht in Vietnam is 20,560MW. Op dit moment heeft Vietnam al 13,694MW geïnstalleerd, goed voor 70% van het potentieel. Volgens het Electricity Development Plan zal Vietnam de doelstelling voor de ontwikkeling van waterkracht in 2030 hebben bereikt. In 2013 zorgde waterkracht voor 40% van de totale productie van elektriciteit in Vietnam. Dit stabiele aanbod van elektriciteit heeft bijgedragen aan de groei van de Vietnamese economie. De Wereldbank en de Aziatische Ontwikkelingsbank hebben hun waardering uitgesproken voor de inspanningen van Vietnam en het succes van het electrificatieprogramma. Op dit moment is meer dan 90% van het landelijk gebied aangesloten op het nationale elektriciteitsnet. Dit heeft bijgedragen aan de bevordering van de sociaaleconomische ontwikkeling en de verbetering van de dienstverlening en de infrastructuur van rurale gebieden. De bevolking van Vietnam draagt het succes van waterkracht en de elektriciteit sector met trots (zie hoofdstuk 2).

De voordelen van waterkracht wegen echter niet op tegen de negatieve effecten op milieu en maatschappij. De bouw van honderden grootschalige hydro-elektrische dammen heeft een enorm verlies van natuurlijke bossen en biodiversiteit veroorzaakt. Honderdduizend hectare aan vruchtbaar landbouwgebied zijn onder water verdwenen in de stuwmeren van deze dammen. Tevens zijn er door de bouw van hydro-elektrische dammen veel conflicten over waterbronnen ontstaan. Dient het gebruik van water om elektriciteit op te wekken of om in de behoeften aan water te voorzien van mensen stroomafwaarts voor de landbouw, transport en drinkwater? Van verschillende dammen is de bouw mislukt, waardoor er veel schade is toegebracht aan bezittingen en land van mensen in stroomafwaarts gelegen gebieden. Aardbevingen hebben zich voorgedaan in de buurt van deze grootschalige dammen waardoor het leven van de omwonenden ernstig is bedreigd. Tot op heden zijn ongeveer 200.000 mensen verplaatst voor de bouw van hydro-elektrische dammen, waarvan meer dan 90% tot etnische minderheden behoort. De meerderheid van verplaatste mensen hebben geen stabiel leven na hervestiging en ondervinden toegenomen moeite om in hun levensonderhoud te voorzien. In de praktijk zijn er zeer weinig voorbeelden van succesvolle hervestiging samenhangend met de bouw van hydro-elektrische dammen in Vietnam (zie hoofdstuk 2, 3, 4 en 5).

Er zijn verschillende onderliggende oorzaken voor de bovengenoemde negatieve gevolgen voor milieu en maatschappij. Met betrekking tot grondwervingsbeleid, compensatie, verplaatsing en hervestiging zijn in Vietnam belangrijke verbeteringen gemaakt, maar de situatie is nog steeds niet wat het moet zijn. Daarbij is er een grote kloof tussen beleid en uitvoeringspraktijk. De implementatie van dit beleid is op lokaal niveau slecht, omdat lokale overheden weinig verantwoordelijkheid nemen en de capaciteit van de instanties die het beleid moeten uitvoeren erg zwak is. In de meeste gevallen wordt een beleidsbeslissing door de lokale en hogere overheden gewoonweg opgelegd aan de lokale bevolking. Belanghebbenden die niet tot de investeerders of lokale overheden behoren worden onvoldoende betrokken in het besluitvormingsproces. De mensen wiens land wordt toegeëigend voor de ontwikkeling van waterkracht hebben niet voldoende rechten en mogelijkheden om te participeren in het besluitvormingsproces met betrekking tot de landverwerving, compensatie, verplaatsing en hervestiging. De meeste beslissingen worden genomen door lokale overheden en investeerders en de stem van de lokale bevolking wordt niet als belangrijk beschouwd. Zo wordt de prijs ter compensatie van land en andere bezittingen elk jaar bepaald door de Provinciale Volkscomité's; deze prijs wordt formeel vastgelegd, waardoor diegenen die land verliezen niet kunnen onderhandelen met investeerders om compensatie tegen marktprijzen to ontvangen.

Bovendien wordt het herhuisvestingsgebied door investeerders en de lokale overheid bepaald, waardoor ontheemde gemeenschappen geen invloed kunnen uitoefenen op de selectie van een geschikte locatie.

Daarnaast is de verantwoordelijk die investeerders nemen ten aanzien van compensatie van beperkte duur, slechts voor een periode van 1 tot 2 jaar na verplaatsing. Na deze periode hebben investeerders geen enkele verplichting om de voordelen die zij behalen uit de waterkrachtdammen met de verplaatste gemeenschap te delen. Zo krijgen deze verplaatste mensen 1 jaar lang gesubsidieerde electriciteir, maar moeten ze daarna betalen voor elektriciteit. Dit is paradoxaal omdat de verplaatste gemeenschap hun land en bezittingen hebben opgeofferd voor de bouw van de stuwdam, maar dan niet mee kunnen profiteren van de vruchten van dat offer. . Daarnaast hebben investeerders en lokale overheden weinig ervaring met dit beleid om verplaatste mensen te ondersteunen in de wederopbouw van hun leven en levensonderhoud. In de meeste gevallen hebben verplaatste gemeenschappen niet voldoende land voor productie en voor werk, waardoor zij veelal geconfronteerd worden met een toegenomen bestaansonzekerheid. Als gevolg nemen ze hun toevlucht tot verouderde en illegale *slash-and-burn* landbouwmethoden en vernietigen zo de bossen, terwijl sommigen terugkeren van hun hervestigingslocatie naar hun gebied van oorsprong om de resterende bossen en land te bewerken. Dit levert conflicten op tussen verplaatste mensen, investeerders, lokale autoriteiten en de lokale bevolking (zie de hoofdstukken 3, 4 en 5).

De negatieve effecten van waterkracht op het milieu en de samenleving hebben geleid tot een beweging tegen de bouw van hydro-elektrische dammen door een netwerk van Non-Goevernementele Organisaties (NGO) in Vietnam en onder de verplaatste gemeenschappen. Dit netwerk heeft zich sterk georganiseerd en is hierdoor in kracht toegenomen. Als gevolg hebben zij aanzienlijke invloed op de ontwikkeling van het waterkrachtbeleid van de Vietnamese regering. Daarnaast hebben deze NGO's een belangrijke plaats gevonden in de Vietnamese samenleving en is de Vietnamese regering toleranter geworden ten aanzien van de activiteiten die zij ondernemen. Tegenwoordig worden NGO's beschermd door wetgeving. Zij hebben hun activiteiten naar verschillende gebieden en velden uitgebreid dankzij hun geslaagde acties tegen de bouw van stuwdammen. Naast Vietnamese NGO's hebben ook internationale NGO's deelgenomen aan de verbetering van het beleid ten aanzien van '*benefit sharing*' uit waterkracht-investeringen, door de Vietnamese overheid en lokale overheden te stimuleren om het model van '*payment for forest environment service*' (FPES) voor de bescherming van bossen te implementeren. Tot op heden hebben vele geherhuisveste gemeenschappen geprofiteerd van dit model. Door het succes van dit beleid heeft de regering de intentie om het model op een nationale schaal uit te

197

breiden. Als gevolg daarvan heeft Vietnam het eerste beleid voor de betaling voor bos,- en milieudiensten in Zuidoost-Azië (zie hoofdstuk 6).

De resultaten van deze studie wijzen erop dat de bouw van honderden stuwdammen veel negatieve gevolgen voor het milieu en de maatschappij heeft veroorzaakt, vooral voor verplaatste gemeenschappen. De ervaringen van Vietnam laten zien dat bij de bouw van hydro-elektrische dammen, van kleinschalige tot grootschalige projecten, in de meeste gevallen zich problemen voordoen. Deze problemen blijven zich terugkomen en hebben een nasleep van vele jaren omdat er nog geen oplossing is gevonden om ze efficiënt op te lossen. Met andere woorden, kwesties gerelateerd aan de ontwikkeling van waterkracht, zoals grondverwerving, schadevergoeding, verplaatsing, hervestiging en herstel van levensonderhoud voor verplaatste gemeenschappen, zijn problematisch en kunnen met de gevolgde aanpak niet succesvol worden opgelost.

De regering van Vietnam, investeerders, en de maatschappij hebben onvoldoende capaciteit en effectieve oplossingen om bestaande problemen gerelateerd aan de bouw van hydro-elektrische dammen op te lossen. Als de regering van Vietnam de ontwikkeling van waterkracht blijft stimuleren, zal de last voor het milieu en de samenleving blijven toenemen en zal het moeilijker worden om deze problemen op te lossen. Waterkracht ontwikkeling kan rechtvaardig en duurzaam zijn wanneer het voordelen voortbrengt die worden gedeeld met verplaatste mensen en worden ingezet om de gevolgen die ze toebrengen aan het milieu en de samenleving te herstellen.

198

TÓM TẮT

Những nghịch lý trong phát triển thủy điện bền vững ở Việt Nam: từ di dân do xây dựng đập thủy điện đến phát triển bền vững?

Thủy điện là một trong những vấn đề gây tranh cãi lớn nhất ở Việt Nam trong thập niên gần đây. Vấn đề gây tranh cãi này liên quan đến những hậu quả mà thủy điện đã và đang gây ra cho môi trường và những người bị dời để xây dựng thủy điện, đặc biệt đa số những người bị di dời là đồng bào dân tộc thiểu số ở các vùng núi cao. Do đó, mục tiêu của cuốn sách này là giải thích rõ hơn về những tranh cãi xung quanh về đề thủy điện ở Việt Nam nhằm đưa ra những gợi ý chính sách cho phát triển công bằng và bền vững. Cuốn sách này tập trung vào việc phân tích các vấn đề nổi bật như thu hồi đất, bồi thường thiệt hại, di dân, tái định cư, hỗ trợ phát triển kinh tế, chia sẻ lợi ích từ phát triển thủy điện, trong đó nhấn mạnh thêm phân tích vai trò của các bên liên quan trong việc đưa ra quyết định phát triển thủy điện ở Việt Nam để thấy rõ hơn mô hình quản trị phát triển thủy điện nhằm giải thích nguyên nhân cho các vấn đề đang gây tranh cãi của thủy điện ở Việt Nam. Nghiên cứu này được tiến hành từ năm 2010 đến 2014, tuy nhiên số liệu và kết quả của các nghiên cứu trước đó cũng được sử dụng để giải thích rõ hơn xu hướng của các vấn đề tranh cãi. Phương pháp nghiên cứu định tính và định lượng đã được áp dụng để thu thập số liệu và phân tích các vấn đề nghiên cứu. Ý kiến của nhiều bên liên quan đến phát triển thủy điện ở các cấp khác nhau được thu thập và phân tích để hiểu rõ hơn vai trò và mức độ ảnh hưởng của các bên liên quan đến quá trình đưa ra các quyết định liên quan đến việc xây dựng thủy điện và phân chia lợi ích tạo ra từ thủy điện ở Việt Nam.

Từ khi độc lập năm 1945 đến nay, Việt Nam luôn coi thủy điện là một trong những chiến lược quan trọng nhất để thúc đẩy công nghiệp hóa và hiện đại hóa đất nước. Tiềm năng thủy điện của Việt Nam là 20,560MW. Hiện nay, Việt nam đã lắp đặt được công suất khai thác đến 13,694MW, chiếm 70% tiềm năng thủy điện. Theo kế hoạch đến năm 2030, Việt Nam cơ bản hoàn thành kế hoạch phát triển thủy điện. Vào năm 2013, thủy điện chiếm khoảng 40% tổng sản lượng điện của cả nước và góp phần vào cung cấp đủ điện cho Việt Nam. Nguồn điện ổn định đã góp phần cho sự tăng trưởng kinh tế Việt Nam trong những năm qua khá ổn định, tỷ trọng các ngành công nghiệp và dịch vụ tăng khá mạnh. Một trong những thành công được đánh giá cao bởi Ngân hàng thế giới và Ngân hàng phát triển Châu Á là tỷ lệ điện khí hóa nông thôn cao hơn 90%, hầu hết các vùng miền của đã được kết nối vào mạng lưới điện quốc gia. Điều này đã góp phần thúc đẩy phát triển kinh tế xã hội và cơ sở hạ tầng cho các vùng nông thôn. Những thành công này là niềm tự hào của người dân Việt Nam về thủy điện và ngành điện (xem chương 2).

Tuy nhiên, những lợi mà thủy điện không thể so sánh với những ảnh hưởng tiêu cực mà thủy điện gây ra cho môi trường và xã hội. Việc xây dựng hàng trăm thủy điện quy mô lớn đã làm mất diện tích lớn rừng tự nhiên và đa dạng sinh học. Những vùng đất màu mỡ cho sản xuất nông nghiệp bị ngập trong các đập thủy điện. Đập thủy điện đã gây ra nhiều mâu thuẫn về tài nguyên nước cho các ngành kinh tế khác nhau, mâu thuẫn giữa việc sử dụng nước để sản xuất điện và nhu cầu nước của người dân ở khu vực hạ lưu. Nhiều đập thủy điện bị vỡ đã làm thiệt hại nhiều tài sản và tín mạng của người dân ở phía hạ lưu. Đặc biệt, nhiều vùng gần đập thủy điện đã xảy ra động đất gây hoang mang cho người dân địa phương và người dân tái định cư. Cho đến nay, có khoảng 200 nghìn người đã bị di dời và tái định của để xây dựng đập thủy điện, trong đó hơn 90% là đồng bào dân tộc thiểu số. Đa số người dân tái định cư đều có cuộc sống không ổn định sau khi tái định cư, đời sống ngày càng khó khăn hơn so với trước khi tái định cư. Có rất ít trường hợp tái định cư do thủy điện được đánh là thành công ở Việt Nam (xem chương 2, 3, 4 và 5).

Những hậu quả do thủy điện gây ra được xác định do các nguyên nhân cơ bản sau. Trước hết, chính sách thu hồi đất, bồi thường thiệt hại, di dân và tái định của Việt Nam đã có những cải thiện đáng kể nhưng vẫn chưa hoàn thiện. Thêm vào đó, việc thực hiện các chính sách này ở các cấp địa phương chưa tốt, trách nhiệm của chính quyền địa phương chưa cao, trình độ quản lý còn thấp, năng lực cán bộ thực hiện các chính sách còn yếu. Việc tổ chức thực hiện các chính sách mang tính áp đặt, thiếu sự tham gia của các bên liên quan trọng khác ngoài chủ đầu tư thủy điện và chính quyền địa phương. Đặc biệt là người dân bị thu hồi đất không có quyền tham gia vào việc đưa các quyết định liên quan đến thu hồi đất, đền bù, di dân và tái định cư. Hầu hết các quyết định đều cho chính quyền địa phương và chủ đầu tư thuỷ điện đưa ra và thực hiện, tiếng nói của người dân không được coi trọng khi đưa ra quyết định. Ví dụ, giá đền bù đất đai và tài sản trên đất là do Ủy ban nhân tỉnh quyết định hàng năm. Người dân không thể đàm phán với chủ đầu tư để đền bù theo giá thị trường. Người dân không có cơ hội để tìm nơi tái định thích hợp mà các khu tái định cư được lựa chọn bởi chủ đầu tư thủy điện và chính quyền địa phương. Hầu hết các khu tái định cư đều nằm ở các khu vực khó khăn, đất đai hẹp và chất lượng đất thấp. Chủ đầu tư chỉ có trách nhiệm đền bù và hỗ trợ cho người dân tái định cư trong vòng 1 – 2 năm, sau đó không có trách nhiệm chia sẽ lợi ích từ việc xây dựng đập thủy điện. Chẳng hạn, người dân tái định cư được trợ cấp điện trong vòng 1 năm, nhưng sau đó họ phải trả tiền điện. Đây là một nghịch lý, vì người dân tái định cư đã hy sinh tài sản và sức khỏe của họ để xây dựng đập thủy điện, nhưng họ lại không được sử dụng điện miễn phí. Ngoài ra, chủ đầu tư và chính quyền địa phương không có các chính sách lâu dài để hỗ trợ người dân tái định cư khôi phục và phát triển sinh kế. Do đó, hầu hết người dân tái định cư không có đủ đất sản xuất, việc làm và hậu quả là họ phải đối mặt với đời sống khó khăn sau khi tái định cư. Hậu quả là họ phải phá rừng để sản xuất, phải quay về nơi ở cũ gần đập thủy điện để khai thác rừng và sản xuất. Điều này gây nên nhiều mâu thuẫn giữa người dân tái định cư với chủ đầu tư thủy điện, chính quyền địa phương và người dân nơi tái định cư (xem chương 3, 4, và 5).

Các tác động tiêu cực của thủy điện đến môi trường và xã hội đã dẫn đến phong trào chống lại việc xây dựng thủy điện của mạng lưới tổ chức phi chính phủ ở Việt Nam và người dân tái định cư. Mạng lưới tổ chức phi chính phủ của Việt Nam ngày càng vững mạnh hơn và tiếng nói của họ đã có ảnh hưởng mạnh mẽ đến việc điều chỉnh chính sách phát triển thủy điện của chính phủ Việt Nam. Ngoài ra, thông qua các phòng trào này các tổ chức phi chính phủ đã tìm được vị trí quan trọng trong xã hội Việt Nam và được chính phủ thừa nhận bằng các văn bản pháp lý. Cho đến nay, các tổ chức phi chính phủ đã mở rộng địa bàn và lĩnh vực hoạt động sau những thành công từ việc can thiệp đến chính sách phát triển thủy điện. Ngoài ra, các tổ chức phi chính phủ quốc tế cũng đã tham gia vào việc cải thiện chính sách chia sẽ lợi ích từ đầu tư thủy điện bằng cách thúc đẩy chính phủ Việt nam và chính quyền địa phương thực hiện các mô hình thu phí bảo vệ môi trường rừng và chi trả cho người bảo vệ rừng cho các hồ thủy điện, trong đó nhiều cộng đồng tái định cư đã được hưởng lợi từ mô hình này. Sự thành công của các mô hình thí điểm về chia sẽ lợi ích đã thúc đẩy chính phủ nhân rộng mô hình trên phạm vi toàn quốc và trở thành chính sách chi trả dịch vụ môi trường rừng đầu tiên ở Đông Nam Á. Sự tham gia của các tổ chức phi chính phủ vào việc cải thiện chính sách phát triển thủy điện đã đem lại những ảnh hưởng mạnh mẽ đến việc điều chỉnh chính sách phát triển thủy điện ở Việt Nam (xem chương 6).

Kết quả nghiên cứu phản ảnh rằng việc xây dựng hàng trăm thủy điện đã gây ra rất nhiều hậu quả tiêu cực cho môi trường và xã hội, đặc biệt đối với người dân bị di dời và tái định cư. Thực tiễn cho thấy, các vấn đề diễn ra ở hầu hết các dự án xây dựng đập thủy điện, ở hầu hết các vùng miền, với các đập thủy điện quy mô khác nhau từ thủy điện quy mô lớn đến quy mô nhỏ. Các vấn đề đó đã diễn ra và kéo dài nhiều năm và không có giải pháp để giải quyết một cách hiệu quả. Nói cách khác là các vấn đề liên quan đến phát triển thủy điện như thu hồi đất, bồi thường thiệt hại, di dân và tái định cư, khôi phục và phát triển sinh kế cho người dân tái định cư là cực kỳ khó khăn và không thể giải quyết thành công. Như vậy, khi chính phủ, chủ đầu tư, và toàn xã hội không có đủ năng lực và giải pháp để giải quyết vấn đề thì tốt nhất là

201

không nên tiếp tục đầu tư phát triển thủy điện để gây ra nhiều hậu quả hơn nữa trong tương lai. Thủy điện phát triển bền vững và công bằng chỉ khi lợi ích mà nó tạo ra được phân bổ hợp lý để khắc phục những hậu quả mà nó gây ra cho môi trường và xã hội một cách hiệu quả, đặc biệt phải có cơ chế rõ ràng cho những người bị di dời để xây dựng thủy điện được đảm bảo hưởng lợi chính đáng từ việc xây dựng thủy điện.

CIRRICULUM VITAE

Ty Pham Huu is currently a lecturer in the Faculty of Land Resources and Agricultural Environment, Hue University of Agriculture and Forestry (HUAF), Vietnam, a position he has held since 2001. He completed a Bachelor's of Land Management (1996 to 2000) at HUAF. Between 2006 and 2008, he completed a Master's of GIS and Remote Sensing at Dalhousie University, Canada. He then returned to work at HUAF, where he carried out several studies related to development-induced displacement and resettlement. He received funding for his research from IDRC and Nanyang University, RDViet project, and Vietnamese Ministry of Training and Education to study GIS and remote sensing, displacement, resettlement, and livelihood impacts due to hydropower dam construction. In mid-2010, he began his PhD research at the International Development Studies (IDS), Department of Human Geography and Planning, Faculty of Geosciences, Utrecht University. His PhD study is funded by the Netherlands Organization for International Cooperation in Higher Education (Nuffic) and Vietnam International Education Development, Ministry of Education and Training (VIED). Furthermore, his research is incorporated into the IS Academy on Land Governance for Equitable and Sustainable Development (LANDac) programme.

During his PhD research, he supervised four IDS Master's students to successfully complete internships in Vietnam. He also delivered two lectures at Utrecht Summer Schools on Land Governance for development in 2010 and 2012. Additionally, he presented various papers at national and international conferences on hydropower dam development in Vietnam. He acted as a peer-reviewer for papers for World Bank Conference 2014 on Land and Poverty. More importantly, he has published several peer-reviewed articles and one book chapter as well as one newsletter. Three chapters in this book are also submitted to peer-reviewed journals to be published. The published papers are listed as follows:

1. Ty, P. H., Van Westen, A. C. M., & Zoomers, A. (2013). Compensation and Resettlement Policies after Compulsory Land Acquisition for Hydropower Development in Vietnam: Policy and Practice. *Land, 2*(4), 678-704.

2. Ty, P. H., Phuc, N. Q., & and Westen, G. v. (2014). Vietnam in the debate on land grabbing: Conversion of agricultural land for urban expansion and hydropower development. In Mayke Kaag and Annelies Zoomers (Ed.), *The global land grab: Behind the hype* (pp. 135-151). London and New York: ZED

203

3. Singer, J., Ty, P. H., & Hai, H. (2014). Broadening stakeholder participation to improve outcomes for dam-forced resettlement in Vietnam. *Water Resources and Rural Development.*

4. Ty, P.H. & Hien, N.T. (2012). Lessons learnt from participatory intervention on land reallocation to displaced households for livelihood restoration in Bo Hon village, Binh Thanh Commune, Thua Thien Hue province, Vietnam. *Vietnam Journal of Ecological Economics,*

5. Ty, P. H., Tu, T. N., & van Westen, G. Food security and energy development in Vietnam. http://www.iias.asia/sites/default/files/IIAS_NL58_2425.pdf